Prime Curios!

Companion Website

http://primes.utm.edu/curios/

Prime Curios!

The Dictionary *of* Prime Number Trivia

Chris K. Caldwell
University of Tennessee at Martin
Martin, TN 38238
caldwell@utm.edu

and

G. L. Honaker, Jr.
Bristol Virginia Public Schools
Bristol, VA 24201
honak3r@gmail.com

CreateSpace (September 2009)

About the Covers

Front cover: Eisenstein primes (see Figure 49 on page 157). The primes are colored red if their norm ($a^2 - ab + b^2$) is an integer squared, otherwise they are colored based on the value of the norm modulo nine. (Cover design by Patrick Capelle and G. L. Honaker, Jr.; image by Chris K. Caldwell.)

Back cover: Top, Moser's circle problem (page 36); left, speed limit in Trenton, Tennessee (page 36); center, Collatz conjecture (pages 91 and 92); bottom-left, a large odd prime that is also "even" (page 1), the prime 379009 (page 177), and paths for the second Central Delannoy prime (Figure 55, page 176). (Images by Chris K. Caldwell; technical assistance by Craig Ingram.)

Credits appear on page 305, which is considered an extension of the copyright page.

CreateSpace

ISBN–10: 1-448-65170-0
ISBN–13: 978-1-448-65170-2

This book is dedicated to those of you
that have enthusiastically supported the
Prime Curios! website.

*Mathematicians have tried in vain to this day to discover
some order in the sequence of prime numbers, and we have
reason to believe that it is a mystery into which the
human mind will never penetrate.*

Leonhard Euler (1707–1783)

Preface

PRIME NUMBERS are those integers *greater than one* which are only divisible by themselves and one, such as 2, 3, 5, 7, 11, and 13. There are numerous books that study the theory of the primes, but here our goal is altogether different: to gather prime number trivia. Short pithy statements about primes which we call Prime Curios!

This book is a labor of love. Here we present the very mathematical alongside the non-mathematical, the coincidental mixed with the deeply significant. Just flip the pages and read a few, in order or at random, to get a feel for what we have collected. If one is too difficult, just move on to the next. This is not a textbook, just a collection of trivia.

For years we have collected prime curios at our website. This edition finally gave us the chance to select the best of these, to expand them and present them as an illustrated dictionary. Enjoy this book, share it with others, then come to our website and add new entries of your own.

Why primes? Prime numbers are the bricks and mortar that numbers are built out of. If you want to understand an integer's properties, you start with its prime factorization. Time to start reading!

The authors gratefully acknowledge the generous assistance received from Patrick Capelle, Patrick De Geest, Shyam Sunder Gupta, Enoch Haga, Mike Keith, Jud McCranie, Carlos Rivera, and the many others who have supported this collection. We especially appreciate the efforts and suggestions from Naomi Caldwell, Stephanie Kolitsch, and Landon Curt Noll. Because of the dedicated work from these wonderful colleagues, the book is far better than it otherwise would have been.

CHRIS K. CALDWELL
G. L. HONAKER, JR.
September 2009

Notes, Symbols and Notations

In this book, the term 'number' means positive integer and all numbers will be written in base ten unless otherwise stated. A number "turned upside down" means to rotate the number 180 degrees about an axis perpendicular to the plane on which the number is written.

We only include curios about numbers which themselves are prime, so the numbers used as entry headings are all primes. All curios listed as smallest known, largest known, and only known, are so as of the date of publication.

The names in brackets at the end of most curios, e.g., [McCranie], are usually the surnames of the persons who submitted those curios, and on occasion, the names of the persons who first discovered the curio(s).

Table 1. Symbols and Notation

symbol	meaning: example
p_i	the ith prime: $p_7 = 17$
$\pi(x)$	prime counting function: $\pi(100) = 25$
$\log x$	the natural logarithm: $\log e = 1$
$n!$	n factorial: $5! = 5 \cdot 4 \cdot 3 \cdot 2 \cdot 1 = 120$
$n!!$	iterated factorial $(n!)!$: $3!! = 6! = 720$
$n!_j$	n multifactorial: $10!_3 = 10 \cdot 7 \cdot 4 \cdot 1$
F_n	Fermat number: $2^{2^n} + 1$: $F_3 = 2^8 + 1 = 257$
$\mathrm{fib}(n)$	Fibonacci number: $\mathrm{fib}(n+1) = \mathrm{fib}(n) + \mathrm{fib}(n-1)$
$M(n)$	Mersenne number: $2^n - 1$: $M(7) = 2^7 - 1 = 127$
$n\#$	n primorial (prime-factorial): $10\# = 7 \cdot 5 \cdot 3 \cdot 2$
digit_i	subscript (repetition) operator: $1_3(31)_2 = 1113131$
$\lfloor x \rfloor$	floor function (round down): $\lfloor \pi \rfloor = 3$, $\lfloor -\pi \rfloor = -4$

Contents

List of Tables

Introduction

Curiosity is one of the permanent
and certain characteristics
of a vigorous mind.

Samuel Johnson (1709–1784)

What is this book?

THIS IS A DICTIONARY of prime number trivia—an eclectic collage of miscellaneous facts. A few of these tidbits have deep mathematical significance, but many are simple observations and often require no mathematics. For example, in what year did England make it illegal to jail a jury for returning the "wrong" decision? What was Jenny's phone number in Tommy Tutone's hit song? What is the highest number of votes a candidate received for the U.S. Presidency while incarcerated? The answers are all prime, and in this book.

Other results are quasi-mathematical, such as those having to do with the shape or representation of a number. Consider the prime 18181. This number is the same forwards, backwards, and even upside down—do you know how many of these primes there are? Can a prime be small and even, and at the same time, large and odd? Take a careful look at the 216-digit prime to the right (read its 216 digits across, then down).

```
700000000000000007
000000222222000000
000022222222220000
000222000002222000
000000000022220000
000000000222200000
000000002222000000
000000222200000000
000022220000000000
000222222222222000
000222222222222000
700000000000000003
```

Have you ever counted how many words could be made by rearranging the letters of the word `stop`? What about the number of primes that can be formed by rearranging subsets of the digits of 13679? Or if you scan the first million digits of π (3.14159...), what is the largest prime you find? The smallest prime you do not find? Again, all of these prime questions are answered in this book.

Finally, we attempt to provide curios for all readers at all levels; so a few of our curios are truly and deeply mathematical. For example, in our entry for the prime 79, we meet the unbelievably large Skewes' number as the upper bound on x below.

$$li(x) < \pi(x) \qquad \text{for some } x \text{ with} \qquad x < e^{e^{e^{79}}}$$

We make some attempt to explain these things, but since this book is written to entertain, not to lecture, please feel free to just move on past those you do not understand.

Why this book?

For one author, primes are an area of research, for the other, a passion; but both of us love recreational mathematics. There are quite a few books of number trivia, but none have focused on just the primes. There are also many excellent books and websites addressing the theory of primes (see our sections "Prime Books" and "Prime Sites"), but this one is just for the pleasure of it.

It is that joy, that pleasure, which is the heart and soul of the science of numbers: number theory. Yet out of idle ideas often comes real mathematics. For example, Stanislaw Ulam, while sitting bored in a meeting, started writing the numbers in an array, beginning with one and then "spiraling" as in Figure 1. When he marked the primes, there seemed to be lines of primes (see Figure 2). These lines represent consecutive values of quadratic functions. Ulam's doodled spiral appeared on the cover of *Scientific American* the following year (March 1964), and still regularly generates research papers.

64	63	62	61	60	59	58	57
37	36	35	34	33	32	31	56
38	17	16	15	14	13	30	55
39	18	5	4	3	12	29	54
40	19	6	1	2	11	28	53
41	20	7	8	9	10	27	52
42	21	22	23	24	25	26	51
43	44	45	46	47	48	49	50

Figure 1. Start of Ulam's Spiral

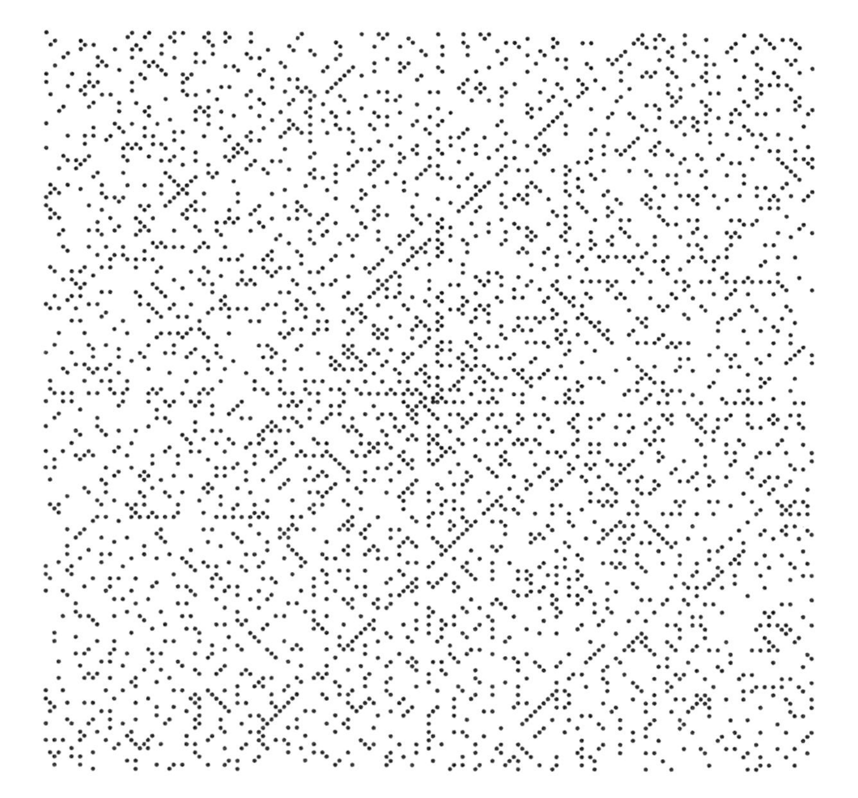

Figure 2. Ulam Spiral

In this work you will find the significant lying beside the mundane, the serious by the silly, all with only you to decide how to categorize them.

How to use this book

The heart of this book is the next chapter "Prime Curios." It is a list of 2151 curios about 1095 prime numbers recorded in dictionary style—the numbers at the top of the pages are the least and largest numbers on those pages. These entries are *only* about prime numbers. After many of the entries there is a name in square brackets, e.g., [Gauss]. This usually refers to the person who suggested it to us, but other times it is the person the contributor chose to credit. We have collected an index of these contributors in the back.

While you read this book you may run into unfamiliar terms. Many are defined in the glossary, and more in the curios themselves. To help you find these definitions we included a subject index in which we boldface the key page numbers.

To wrap things up we end with a chapter on the $100,000 prime—the largest prime known to man (as well as a short history of prime number records), and of course, a list of primes.

No book such as this can be a finished work. Records are regularly broken. Errors might have slipped by our repeated proofing. So come visit the book's website: `http://primecurios.com`.

Two old friends

Ever since the study of primes began, a key question has been "how many are there?" About 2300 years ago Euclid showed that there are infinitely many, so we ask "how many primes are less than (or equal to) x?" Mathematicians love short expressions, so we use the symbol $\pi(x)$ for the answer to this question (π for $\pi\rho\hat{\omega}\tau o\varsigma$, the name Euclid used for primes).

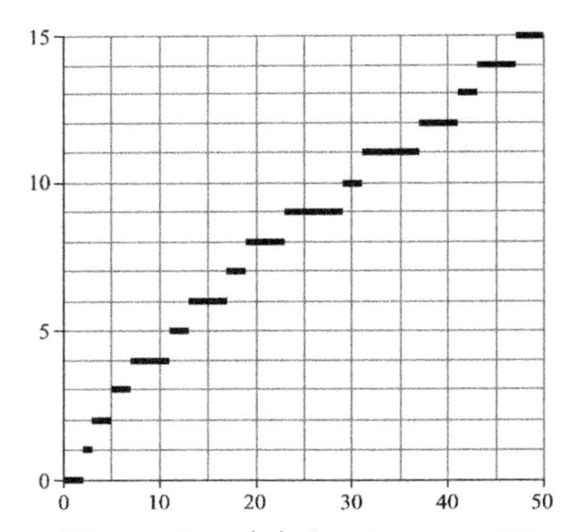

Figure 3. $\pi(x)$ for $0 \le x \le 50$

Since the first few primes are 2, 3, 5, and 7, $\pi(1) = 0$, $\pi(2) = 1$, $\pi(9) = 4$, and (we cannot help adding) $\pi(\pi) = 2$. The first few values of these are graphed in Figure 3. We mention this function because it may be new to you and will show up many times in this book (in fact it already has when we discussed Skewes' number above). Table 10 (page 179) has a list of key $\pi(x)$ values, and on page 57 we have a larger graph.

Our second old friend is modular arithmetic (mod). It is essentially arithmetic using just remainders. For example, you might see $a \equiv 1 \pmod 6$ (or "a is 1 modulo 6"). This just means that when you divide a by 6 the remainder is 1. Did you know that every prime p greater than 3 leaves a remainder of 1 or 5 when you divide it by 6? Using this notion we can write $p \equiv \pm 1 \pmod 6$ (because $5 = 6 - 1$, so both 5 and -1 leave the same remainder when divided by 6). Let us list two more powerful examples of this tool.

> **Fermat's Theorem:** If p is any prime, and a any number not divisible by p, then $a^{p-1} \equiv 1 \pmod p$.
>
> **Wilson's Theorem:** $p > 1$ is a prime number if and only if $(p-1)! \equiv -1 \pmod p$.

Modular arithmetic is sometimes called clock arithmetic in primary school. Our clocks report the minutes modulo 60 and the hours modulo 12.

Before you begin perusing the trivia, let us end with one final question: What do the following four infinite expressions equal?

$$\sqrt{2+\sqrt{2+\sqrt{2+\sqrt{2+\sqrt{2+\cdots}}}}}$$

$$\sqrt{3+\sqrt{3-\sqrt{3+\sqrt{3-\sqrt{3+\cdots}}}}}$$

$$\sqrt{7-\sqrt{7+\sqrt{7-\sqrt{7+\sqrt{7-\cdots}}}}}$$

$$\sqrt{19-3\sqrt{19+3\sqrt{19-3\sqrt{19+3\sqrt{\cdots}}}}}$$

(Hint: they are all the same prime.)

Prime Curios!

2

The first prime, and the only even prime. (Does that also make it the "oddest" prime?)

The Pythagoreans considered 2 to be the first feminine number.

De Polignac's Conjecture states that every even number is the difference of 2 consecutive primes in infinitely many ways.

The addition and product of 2 with itself are equal, which gives it a unique arithmetic property among the positive integers.

2! is the only factorial that is prime.

The smallest untouchable number, i.e., an integer that cannot be expressed as the sum of all the **proper divisors** of any positive integer (including the untouchable number itself). The first few are 2, 5, 52, 88, 96, ...

The probability that the greatest prime factor of a random integer n is greater than $\sqrt{n}$ equals the natural **logarithm** of 2. [Schroeppel]

Euler's formula: $V - E + F = 2$. For any convex polyhedron, the number of vertices and faces together is exactly two more than the number of edges.

German "Euler" Stamp

The only "eban" prime, i.e., devoid of the letter 'e' in its English name. [Beedassy]

Fermat's Last Theorem: The equation $x^n + y^n = z^n$ has no solution in positive integers for n greater than 2. [Wiles]

François Viète (1540–1603) expressed π as an infinite product containing only 2 (and its reciprocal $\frac{1}{2}$).

$$\pi = \frac{2}{\sqrt{\frac{1}{2}}\sqrt{\frac{1}{2}+\frac{1}{2}\sqrt{\frac{1}{2}}}\sqrt{\frac{1}{2}+\frac{1}{2}\sqrt{\frac{1}{2}+\frac{1}{2}\sqrt{\frac{1}{2}}}}\cdots}$$

UCLA mathematician and prime number researcher Terence Tao taught himself arithmetic at age 2.

Figure 4. Six-Inch Ruler with Two Marks

It is possible to measure all of the integer distances from one to six on a six-inch ruler with just 2 marks (Figure 4). For example, the distance from the 2 to the right end is four inches.

3

The first odd prime number.

$\pi(3!) = 3$.

Captain Kirk and Spock played chess 3 times on the television series *Star Trek*. Kirk won every game.

The smallest **reflectable prime**.

Reflect: 2 3 5 7

The Italian-born French mathematician
Joseph-Louis Lagrange (1736–1813) spent much of his life working on the 3-Body Problem.

The first in a pair of primes of the form $(p, p + 4)$ called **cousin primes**.

The smallest **Fortunate number**.

Choose a prime number greater than 3. Multiply it by itself and add 14. If the result is divided by 12, then the remainder will always be 3.

Vinogradov's theorem states that every sufficiently large odd integer is a sum of at most 3 primes.

The first lucky prime. Since lucky numbers are lucky enough to repeatedly appear in this book, let's take a moment to define them. Start with the list of natural numbers: 1, 2, 3, ..., and cross out every second number. The second number not crossed out is 3, so we cross out every third number leaving 1, 3, 7, 9, 13, 15, ... The third number left is 7, so we cross out every seventh number—repeat forever. What remains is the sequence of lucky numbers:

$$1, 3, 7, 9, 13, 15, 21, 25, 31, 33, 37, 43, 49, 51, 63, 67, 69, \ldots.$$

S. Ulam (1909–1984) investigated the lucky numbers and found a strong resemblance to primes.

Divisibility test for 3: A number is divisible by 3 if the sum of its digits is divisible by 3. (The same is true for nine.) [Greene]

The smallest **Fermat prime**. (Fermat primes are the primes of the form $2^{2^n} + 1$.)

The "Three-fold Law" is a common tenet held by some Wiccans stating that both the good and the evil that one creates in the world come back to benefit or hurt them—magnified 3 times over.

Racing legend Dale Earnhardt drove the number 3 car for most of his career. (His first car was pink "K-2".)

Octopuses have 3 hearts.

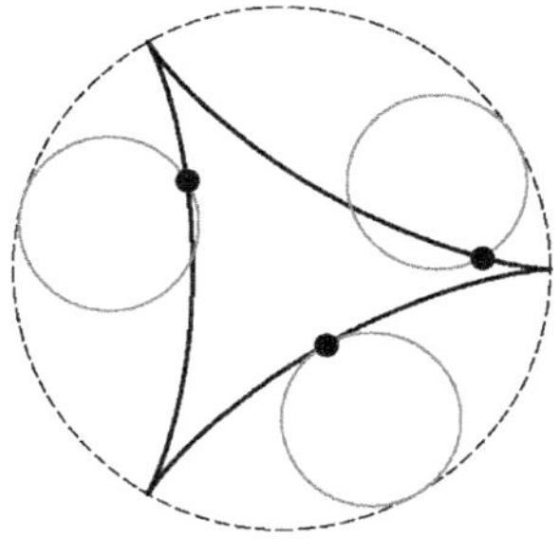

A mark on a small circle, rolling inside one with three times the diameter, traces out a 3-cusped hypocycloid. Euler called it a deltoid because of its resemblance to the Greek letter delta (Δ).

Nicola Tesla (1856–1943), inventor, electrical engineer, and physicist, was obsessed with the number 3. For example, it was not uncommon to see him walk around a block 3 times before entering a building.

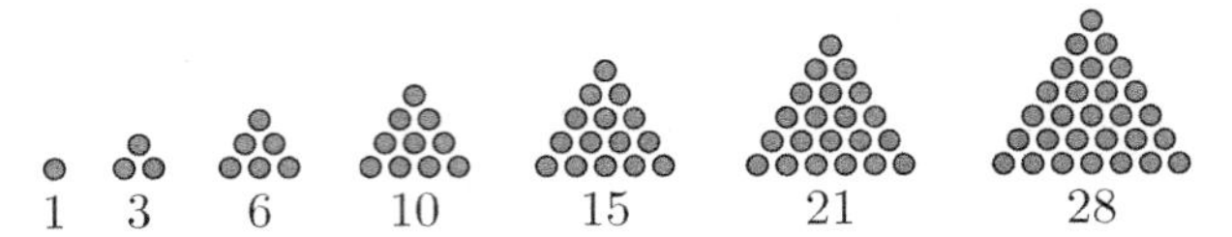

Figure 5. The First Seven Triangular Numbers

Sharkovsky's theorem states that if a continuous real-valued function has a point of period 3 (i.e., $x = f(f(f(x)))$), then it has points of every other period.

If White's chess pieces are on their original squares and Black has only a king on h4, then White can checkmate Black in 3 moves. [Loyd]

The 3-toed sloth reaches sexual maturity at about 3 years of age. [Jinsuk]

The dog-sized *Eohippus* ("dawn horse") had 3 hoofed toes on each hind foot. [Marsh]

The German card game Skat requires at least 3 players. [Luhn]

The largely self-taught Indian genius Srinivasa Ramanujan (1887–1920) noticed that 3 is

$$\sqrt{1+\sqrt{1+2\sqrt{1+3\sqrt{1+4\sqrt{1+\ldots}}}}}.$$

The only **Fermat number** which is also a triangular number. (Triangular numbers are illustrated in Figure 5.) [Gupta]

There are 3 additive primary colors (red, green, and blue) and 3 subtractive primary colors (cyan, magenta, and yellow).

The Pythagoreans considered 3 to be the first masculine number.

The function $n^{\frac{1}{n}}$ achieves its maximum value for integers n at $n = 3$. [Rupinski]

In most jurisdictions, a tablespoon equals 3 teaspoons (but it is 2 teaspoons in Asia and 4 in Australia).

3 is the first **Mersenne prime** (i.e., a prime of the form $2^n - 1$). [Rajh]

A codon is a sequence of three adjacent nucleotides, which codes for an amino acid. [Necula]

NUMB3RS is an American television show that airs on CBS. In one episode called *Prime Suspect*, a young girl's kidnapping is related to her father's work on the **Riemann hypothesis**.

The law of proportions, called "Rule of Three" by the Indian mathematician Brahmagupta (598–668), became a standard of rational thought. For example, Abraham Lincoln wrote that as a young man he "could read, write, and cipher to the Rule of Three." Charles Darwin wrote "I have no faith in anything short of actual measurement and the Rule of Three." Perhaps less known is the fact that they were born on the exact same day (February 12, 1809).

The smallest **triadic prime**. [Capelle]

The terms of the sequence $\frac{3}{2}, \frac{5+7}{2+3}, \frac{7+11+13}{2+3+5}, \frac{11+13+17+19}{2+3+5+7}$, etc., converge to 3 as the primes used approach ∞.

The smallest odd **Fibonacci prime**. It is the only Fibonacci prime with a composite index number: $3 = \text{fib}(4)$.

5

Provably the only prime that is a member of two pairs of **twin primes**. [Pallo]

There are 5 Platonic solids (convex regular polyhedra; Figure 6).

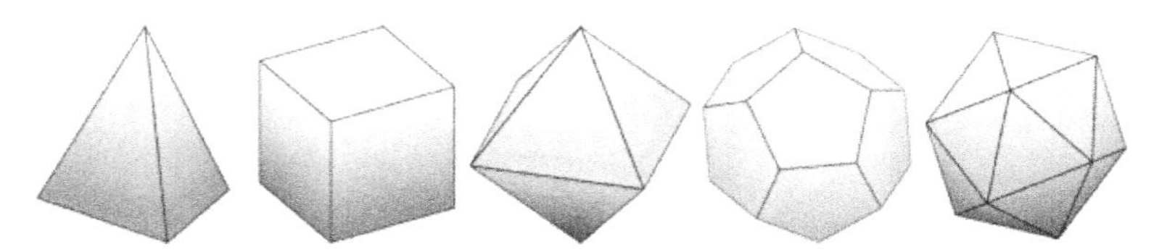

Figure 6. The Five Platonic Solids

Euclid gave 5 postulates of plane geometry. [McCranie]

The smallest prime in the first **sexy prime** pair (5, 11). Prime pairs differing by six are "sexy" because *sex* is the Latin word for six. [Wilson]

The first prime of the form $6n - 1$.

There are no known **Wall-Sun-Sun primes** greater than 5.

The smallest **balanced prime**.

The 5th Mersenne prime is $-1 + 2^3 \cdot 4^5$.

5 is believed to be the only odd untouchable number.

$5 = 3! - 2! + 1!$.

The smallest **Wilson prime**.

A cryptarithm is a type of mathematical puzzle in which most or all of the digits in a mathematical expression are substituted by letters or other symbols. In the case of XZY + XYZ = YZX, the value of Z must equal 5.

The 5th **Fibonacci number**.

$n!$ never ends in 5 zeros. Note that the first 5 terms in the sequence of numbers of zeros that $n!$ never ends in are all prime. The set includes 5, 11, 17, 23, and 29.

The only prime that is the sum of "Siamese twins," i.e., 2 and 3, which are the only pair of primes that are conjoined (have no composite between them). [Gevisier]

A limerick is a light humorous or nonsensical verse of 5 lines that usually has the rhyme scheme *aabba*. [Luhn]

The fewest number of moves for pawn promotion to occur in chess. [Rachlin]

Alan Turing's Erdős number (see page 111). [Croll]

The American 5-cent piece called a "nickel" weighs 5.000 grams. It used to be made of nickel but is now mostly a copper alloy. [Lee]

5 is the 5th digit in the decimal expansion of $\pi = 3.14159....$ [Gupta]

The smallest **safe prime**. [Russo]

The smallest **good prime**. [Russo]

The first 5 open-end aliquot sequences (recursive sequences in which each term is the sum of the proper divisors of the previous term) are the so-called "Lehmer five."

An ace is a military aircraft pilot who has destroyed 5 or more enemy aircraft. [Richthofen]

One of the best-known perfumes, *Chanel $N^{\circ}5$*, was introduced by Gabrielle "Coco" Chanel on May 5, 1921.

Is "abstemious" the only English word which uses all 5 vowels just once in alphabetical order and contains the same number of consonants? [Bown]

The smallest odd prime Thâbit number. Arab mathematician Thâbit ibn Qurra discovered a rule for finding amicable pairs based on these numbers (of the form $3 \cdot 2^n - 1$) before his death in Baghdad in 901.

The masculine marriage number to the Pythagoreans, uniting the first female number and the first male number by addition.

The only temperatures that are prime integers in both Celsius and Fahrenheit are $\pm 5\,^{\circ}$C ($41\,^{\circ}$F and $23\,^{\circ}$F).

The sum of the reciprocals of the primes ($\frac{1}{2} + \frac{1}{3} + \frac{1}{5} + \ldots$) is infinite, but the sum of the reciprocals of the *known* primes is less than 5 *and will always be so!*

Every number can be written as $x^2 + 2y^2 + 7z^2 + 11w^2$, except for 5.

7

The first prime number of the form $6n + 1$. All primes greater than three have the form $6n \pm 1$.

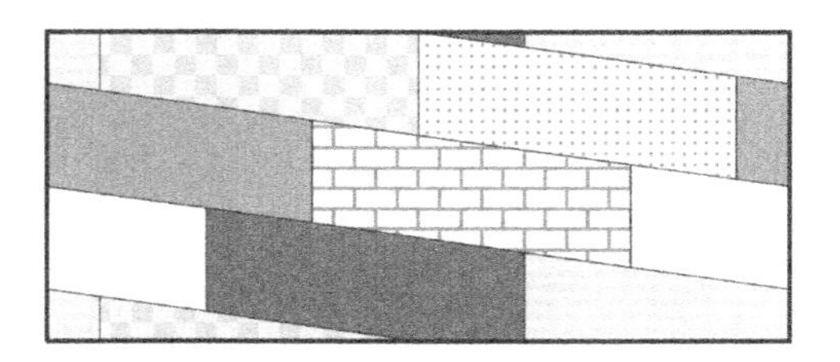

Figure 7. Pattern for a Toroidal Map

Graph theorists have been able to prove that 7 colors are required on a donut-shaped map (i.e., an ordinary one-holed torus) to ensure that no adjacent areas are the same. To create such a map, roll the pattern in Figure 7 into a tube by connecting the top to bottom, and then make it a torus (doughnut) by connecting the two ends together.

There are 7 letters in TUESDAY, which is the only day of the week whose name contains a prime number of letters. [Gupta]

The first 7 digits of 8^9 form a prime. [Kulsha]

$7! - 6! + 5! - 4! + 3! - 2! + 1!$ is prime. [Guy]

Seven is the only odd prime that becomes "even" by deleting a letter. [Beedassy]

"Seventh heaven" is the farthest of the concentric spheres containing the stars in the Muslim and Cabalist systems.

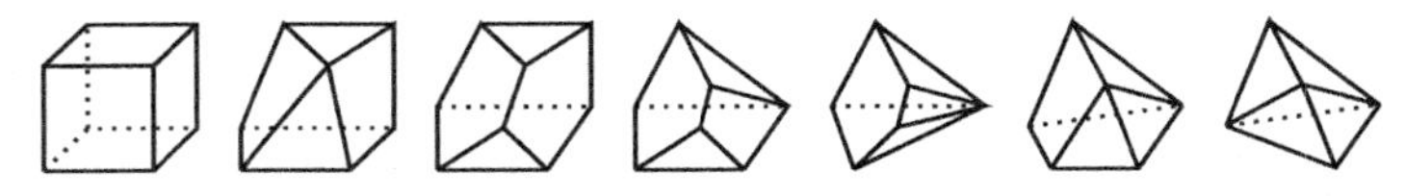

Figure 8. The Seven Basic Hexahedra

There are exactly 7 possible basic shapes for six-sided polyhedra (convex hexahedra; Figure 8).

Pill bugs ("roly-polies") have 7 pairs of legs.

In 1992, Bayer and Diaconis showed that after 7 random riffle shuffles of a deck of 52 cards, every configuration is nearly equally likely.

The Seven Bridges of Königsberg is a famous solved mathematics problem inspired by an actual place (now Kaliningrad, Russia) and situation. [Millington]

The smallest odd full-**period** prime.

The first prime happy number. Replace a number by the sum of the squares of its digits, and repeat the process until the number equals 1, or it loops endlessly in a cycle which does not include 1. Those numbers for which this process ends in 1 are called happy numbers. There are exactly 7 prime happy numbers less than 100. Can you find all 7?

There are 7 indeterminate forms involving 0, 1, and ∞ that arise when evaluating limits:

$$\frac{0}{0}, \quad \frac{\infty}{\infty}, \quad 0 \cdot \infty, \quad 0^0, \quad \infty^0, \quad 1^\infty, \quad \infty - \infty.$$

7 Up (or Seven Up) is a brand of a lemon-lime flavored soft drink.

The Clay Mathematics Institute announced in May 2000 a prize of $1,000,000 for each of the 7 unsolved Millennium Prize Problems. The Poincaré conjecture (essentially the first deep conjecture ever made in topology) has been the only Millennium problem solved thus far.

There are 7 different notes in a standard major scale in music as well as in a standard minor scale. [Obeidin]

Double 7! to get the exact number of minutes in a week (7 days). [Luhn]

There are "7 deadly sins" used in early Christian teachings to educate and protect followers from basic human instincts (pride, envy, gluttony, lust, anger, avarice, and sloth). [Croll]

There were 7 rings of power for the Dwarf-lords in the stories of J.R.R. Tolkien. [Oldenbeuving]

The number of points and lines on the minimal finite projective (or Fano) plane. [Poo Sung]

It is possible to place 7 cigarettes in such a way that each touches the others if the length divided by the diameter of each is greater than or equal to $\frac{7\sqrt{3}}{2}$. [Gardner]

There is a persistent rumor that the philosopher G.W.F. Hegel (1770–1831) provided a logical proof within his doctoral dissertation that there could only be 7 planets in the solar system. [Poo Sung]

In most Hindu marriages the bride follows the groom 7 times around the holy fire, which is called a Saptapadi. [Das]

Uranus' moon Miranda (as viewed from space) has a "7" embedded in the middle of a rectangular corona that appears to have been formed by viscous icy lavas. Uranus is the 7th planet from the Sun.

The bluntnose sevengill shark has 7 long gill slits in front of each pectoral fin. [Jolly]

The number of distinct prime knots containing 7 crossings (Figure 9). [Tait]

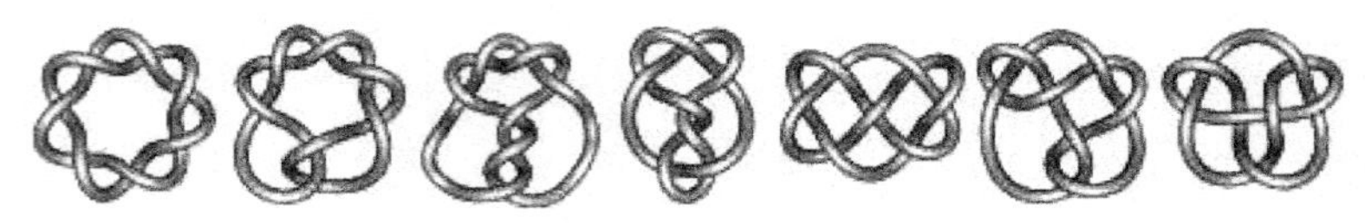

Figure 9. Seven Prime Knots

Kwanzaa (or Kwaanza) is a 7-day festival (December 26 to January 1) celebrated primarily in the United States, honoring African-American heritage. [Karenga]

The only prime that can be the digital root (sum of digits computed recursively until one digit remains) of a perfect square. [Rupinski]

Andrew Wiles spent 7 years working on his solution of Fermat's Last Theorem. [Croll]

In 1995, Ramaré showed that all integers greater than one are the sum of at most 7 prime numbers.

The Japanese word "Subaru" refers to an open star cluster called the Seven Sisters (Pleiades).

The numbers on opposite sides of a standard die always add up to 7.

The number of rank and good ears of corn that came up upon one stalk in Pharaoh's second dream (Genesis 41:5).

Assuming **Goldbach's conjecture**, there is no positive number that can be written as the sum of exactly 7 primes in exactly 7 ways. [Hartley]

Have you ever noticed that the word "indivisibilities" contains 7 i's?

The sum of seven consecutive primes beginning with 7 is seven times the seventh prime. [Post]

The maximum number of eclipses of the Sun and Moon that can occur in any one year. [Byrd]

The smallest **cuban prime**. The name has nothing to do with Cuba the country.

Levy's conjecture (1963) states that all odd numbers greater than or equal to 7 are the sum of a prime plus twice a prime. It was named after Hyman Levy, who was apparently unaware that the conjecture was first stated by Émile Lemoine in 1894. [Capelle]

Early electronic instruments incorporated alphanumeric displays that employed discrete circuit components known as 7-segment light-emitting diodes (LED's). [McAlee]

The largest known number of consecutive primes that can be partitioned into two sets such that the difference of their products is unity: $5 \cdot 11 \cdot 13 - 2 \cdot 3 \cdot 7 \cdot 17 = 1$.

Divisibility test for 7 (or 13): Combine the digits in order into groups of 3 (starting from the right) by alternating them with positive and negative signs. If the result is divisible by 7 (or 13 respectively), then so is the original number. For example, 7 divides 62540982 because 7 divides $+(62) - (540) + (982)$.

11

The only **palindromic prime** with an even number of digits due to the fact that all **palindromes** with an even number of digits are divisible by 11.

11^{11} contains exactly two embedded elevens.

The secret formula for Kentucky Fried Chicken includes 11 herbs and spices. [Sanders]

Sunspot activities seem to follow an 11-year cycle.

Rotakas, spoken in the center of Bougainville Island in the South Pacific, uses only 11 phonemes.

The smallest anti-**Yarborough prime**, i.e., a prime containing only the digits 0 and 1.

Five consecutive powers of 11 produce palindromes.

$$11^0 = 1$$
$$11^1 = 11$$
$$11^2 = 121$$
$$11^3 = 1331$$
$$11^4 = 14641$$

The number of cards including the Significator (the focus card) typically used in a Tarot reading. [Haga]

"Elevenses" is a British and colonial meal that is similar to afternoon tea but eaten around 11 o'clock in the morning.

The first **repunit prime**. The term repunit (coined by A. H. Beiler in 1966) comes from the words *rep*eated and *unit*, so repunits are positive integers in which every digit is 1.

Paul Erdős observed that both $3 \cdot 4$ and $5 \cdot 6 \cdot 7$ are congruent to one modulo 11.

The number of letters in PRIME NUMBER. [Kumar]

Today only 11 lines of Sotades the Obscene of Maronea's works still remain. Most sources credit him with inventing palindromes in Greek-ruled Egypt, back in the 3rd century B.C.

The smallest prime p such that $2^p - 1$ is composite. [Russo]

The original formulation of M-theory was in terms of a (relatively) low-energy effective field theory, called 11-dimensional supergravity.

ElevenSmooth is an online distributed computing project searching for prime factors of $M(3326400)$.

Can you solve the doubly-true alphametic (on the right) by replacing each letter with a different digit to make a valid arithmetic sum?

$$\begin{array}{r} \text{THREE} \\ \text{TWO} \\ \text{ONE} \\ \text{TWO} \\ +\ \text{THREE} \\ \hline \text{ELEVEN} \end{array}$$

A hendecasyllabic is verse written in lines of exactly 11 syllables. [Patterson]

Self-proclaimed psychic Uri Geller (1946–), has spoken repeatedly about the occurrence of two 11's side-by-side. For example, the bizarre attraction some people have for the time 11:11 on digital clocks and watches.

The Works of Charles Babbage, published in London by Pickering and Chatto Publishers, is an 11-volume set. [McAlee]

World War I ended formally at 11 A.M. on the 11th day of the 11th month of the year. Now the date is celebrated as Veteran's Day. [McCranie]

The name for the now dwarf planet Pluto (discovered by Clyde W. Tombaugh in 1930) was proposed by 11-year-old Venetia Burney of

Oxford, England. She is now a retired teacher whose married name is Venetia Phair. [Paddy]

Joseph-Louis Lagrange (1736–1813) is widely regarded as the finest mathematician of the 18th century. He was the first-born of 11 children.

The largest integer that cannot be expressed as a sum of (two or more) distinct primes. [Capelle]

A hendecagon (or undecagon) is an 11-sided polygon. The shape surrounds the portrait on the Susan B. Anthony one-dollar coin. [Patterson]

If n is sufficiently large, then between n and $n + \sqrt{n}$ there exists a number with at most 11 prime factors. [Brun]

Aibohphobia (the fear of palindromes) is palindromic itself and contains 11 letters. [Patterson]

Substance P is an 11-amino acid polypeptide that has been associated with the regulation of stress brought about by failure to find large primes.

The smallest odd **Ramanujan prime**. [Beedassy]

Divisibility test for 11: Combine the digits in order by alternating them with positive and negative signs. If the result is divisible by 11, then so is the original number. For example, 11 divides 90816 because 11 divides $+9 - 0 + 8 - 1 + 6$. [Beedle]

13

There are 13 Archimedean solids.

The smallest **emirp**.

13 is the only prime that can divide two successive integers of the form $n^2 + 3$. [Monzingo]

A "baker's dozen" is a group of 13. Its origin can be traced to a former custom of bakers to add an extra roll as a safeguard against the possibility of twelve weighing light.

$13^2 = 169$ and its **reversal** $31^2 = 961$.

The olive branch on the back of a U.S. one-dollar bill has 13 leaves. [Weirich]

13^2 turned **upside down** is prime.

There is no elliptic curve over the rationals Q having a rational point of order 13. [Mazur and Tate]

The United States flag once had 13 stars and 13 stripes, which represented the 13 original colonies.

ELEVEN + TWO = TWELVE + ONE. Can you solve this doubly-true anagrammatical equation?

Alfred Hitchcock's directorial debut was the film *Number 13*, which was never completed. [Liebert]

$M(13)$ can be expressed as the sum of 13 consecutive primes $(599 + 601 + \ldots + 661)$. [Vrba]

The sum of the remainders when 13 is divided by all the primes up to 13. [Murthy]

The Jewish sage Moses Maimonides established 13 principles of the Jewish faith during the Middle Ages. [Croll]

A concatenation of the first two triangular numbers. [Gupta]

The dice game Yahtzee consists of 13 rounds. [Bailey]

Three planes can cut a donut into a maximum of 13 parts. [Laurv]

A female Virginia opossum usually has 13 nipples: twelve efficiently arranged in an open circle with one in the center. The length of gestation for this curious marsupial is about 13 days. [McCarthy]

$\pi(13) = 1! \cdot 3!$. [Gupta]

Girolamo Cardano (1501–1576) divided cubic equations into 13 types (excluding $x^3 = c$ and equations reducible to quadratics) in his work *Ars Magna*, which was the first Latin treatise devoted solely to algebra. [Poo Sung]

The fear of the number 13 is called triskaidekaphobia.

The term paraskevidekatriaphobia was coined by therapist Dr. Donald Dossey from the Stress Management Center/Phobia Institute in Asheville, North Carolina, and refers to the fear of Friday the

13th. He claims that when you learn how to pronounce the word (*pair-uh-skee-vee-dek-uh-tree-uh-FOH-bee-uh*), you'll be cured of the affliction. [Hammond]

Ironically, 13 is a lucky number (page 9). [Emmert]

The sum of primes up to 13 is equal to the 13th prime. [Gupta]

The 13th of May 2011 will be a "double Friday the 13th," i.e., the sum of the digits of 5/13/2011 equals 13. The next time this will happen in a prime year is 1/13/2141.

The smallest prime that can be expressed as the sum of two primes $(2 + 11)$ and two composites $(4 + 9)$ in only one way. [Gupta]

Patau syndrome, also known as trisomy 13, is the consequence of a rare chromosomal abnormality. [Smith]

A three-digit number abc is divisible by 13 if 13 divides $a + 4b + 3c$. (See also the divisibility test on page 17.)

17

The only prime that is the average of two consecutive Fibonacci numbers. [Honaker]

Theodorus of Cyrene (5th century B.C.) proved that the square root of each of the odd primes up to 17 is irrational. It is not known why he stopped at 17.

All calculus students should know basic trigonometric values such as $\cos\left(\frac{2\pi}{3}\right)$. But have you ever seen Gauss's formula for $\cos\left(\frac{2\pi}{17}\right)$?

$$\cos\left(\frac{2\pi}{17}\right) = -\frac{1}{16} + \frac{1}{16}\sqrt{17} + \frac{1}{16}\sqrt{2\cdot 17 - 2\sqrt{17}} + \frac{1}{8}\sqrt{17 + 3\sqrt{17} - \sqrt{2\cdot 17 - 2\sqrt{17}} - 2\sqrt{2\cdot 17 + 2\sqrt{17}}}$$

The only known prime that is equal to the sum of digits of its cube ($17^3 = 4913$ and $4 + 9 + 1 + 3 = 17$). [Gupta]

The smallest odd number greater than three that cannot be represented as the sum of a prime and twice a nonzero square. [Stern]

A smile uses 17 muscles. :-)

"17-jewel watches" have hard gems at 17 bearing points of friction.

The 17-stringed bass koto (or *jūshichi-gen*, literally "seventeen strings") is used in the album *PRIME NUMBERS* by Neptune/Watanabe. As a matter of fact, all the traditional Japanese instruments played in it contain prime numbers. The other instruments used are the 3-stringed shamisen, the 5-holed shakuhachi (bamboo flute), and the 13-stringed koto.

Frank Bunker Gilbreth (1868–1924), pioneer of modern motion study technique, and his wife Lillian devised a classification scheme to label 17 fundamental hand motions of a worker, which they called *therbligs* ("Gilbreth" spelled backwards with the *th* transposed).

17 is the smallest natural number that is written in French as a compound word: *dix-sept*. [Lefèvre]

"Seventeen" was the original title of The Beatles' song "I Saw Her Standing There." It was written on a Liverpool Institute exercise book. [McCranie]

The smallest prime that is both the sum of a prime number of consecutive composites, and also, the sum of a composite number of consecutive primes: $17 = 8 + 9 = 2 + 3 + 5 + 7$. [Beedassy]

The world's largest caldera is that of Mt. Aso in Kyushu, Japan, which measures 17 miles north to south and 71 miles in circumference. (17 and 71 are emirps.)

The 17-year locust has the longest cycle of development of any known insect. Note that its Latin name (*Cicada septemdecim*) has 17 letters. The genus *Magicicada* is sometimes called the "seventeen-year locust," but they are not locusts; locusts belong to the order Orthoptera. [Dillon]

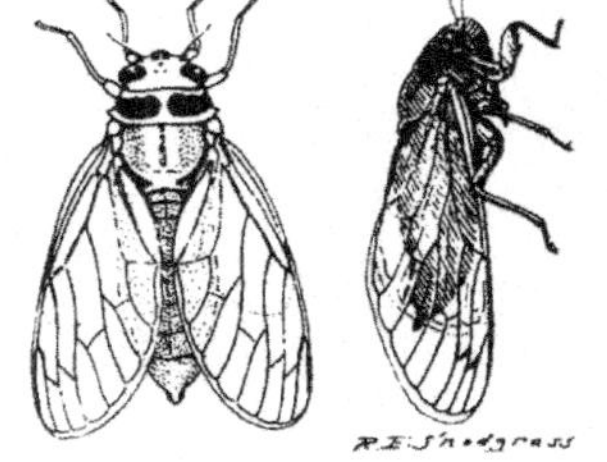

Moderately active people can estimate their daily calorie requirement by multiplying their weight in pounds by 17. [Pierson]

President Bill Clinton's dog Buddy was killed by a vehicle driven by a 17-year-old girl.

"At Seventeen" was a hit song for Janis Ian in 1975. [Litman]

No odd Fibonacci number is divisible by 17. [Honsberger]

17 is the smallest prime sandwiched between two non-squarefree numbers. A number is squarefree if it is not divisible by the square of an integer greater than 1. [Gupta]

Schnizel showed that Goldbach's conjecture is equivalent to saying that every integer greater than 17 is the sum of three distinct primes.

Marin Mersenne's vast correspondence (*Correspondance du Père Marin Mersenne, religieux minime*) fills a total of 17 volumes.

Mersenne (1588–1648)

There are exactly 17 ways to express 17 as the sum of one or more primes. [Rupinski]

"Seventeen or Bust" is a distributed attack on the Sierpiński problem. This problem asks whether $k = 78557$ is the least odd number such that $k \cdot 2^n + 1$ is composite for all $n > 0$ (these k's are called Sierpiński numbers). When the project began in 2002 there were only 17 values of $k < 78557$ that were still in question. [Mendes]

Stalag 17 is a classic film set in a German prisoner of war (POW) camp. [Hartley]

17 Lectures on Fermat Numbers: From Number Theory to Geometry was written in honor of the 400th anniversary of the birth of amateur mathematician and lawyer Pierre de Fermat. (17 is a Fermat prime.)

According to hacker's lore, 17 is described at MIT as "the least random number."

The longest word of prime length in the King James Version of the Bible (the "KJV Bible") is *Chushanrishathaim* with 17 letters. [Opao]

The middle verse in the New Testament is Acts 17:17.

Vietnam was divided along the 17th parallel of latitude.

The "*Encyclopédie*" was published in France with 17 volumes of articles issued from 1751 to 1765. Denis Diderot coedited the monumental work with Jean le Rond d'Alembert (born in 1717). [Marr]

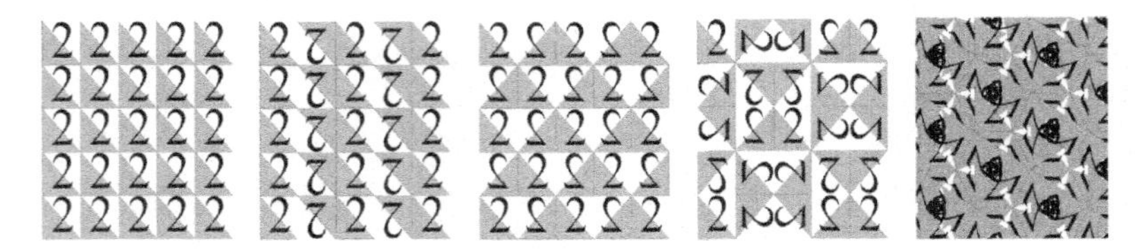

Figure 10. Five (of Seventeen) Wallpaper Symmetries

There are 17 plane symmetry groups, i.e., there are 17 different ways that a wallpaper design can repeat (Figure 10).

17 is the $(1 \cdot 7)$th prime. [Firoozbakht]

The sum of the first 17 composite numbers is prime. [Patterson]

There were 17 episodes of the classic BBC TV series *The Prisoner*. [McCranie]

It is believed (though not proven) that the minimum number of hints in a Sudoku grid that will lead to a unique solution is 17 (see Figure 61 on page 221). [McCranie]

There is an absence of Y-chromosome marker H17 in Polynesian populations. [Brookfield]

The first program to run on a stored-program computer consisted of 17 instructions. It was used to find the largest factor of $2^{30} - 1$. [McCranie]

Stegosaurus had 17 bony plates that were embedded in its back.

There are 17 words in the following quotation: "It is evident that the primes are randomly distributed but, unfortunately, we don't know what 'random' means." R. C. Vaughan (February 1990). [Post]

There are 17 standard-form types of quadratic surfaces (quadrics). [Poo Sung]

The largest known prime that is not the sum of two **semiprimes**. [Manor]

Ramanujan defined 17 Jacobi theta function-like functions which he called "mock theta functions" in his last letter to Hardy.

The sum of digits of the square of 17 is its twin prime. [Silva]

David C. Kelly, a math professor at Hampshire College in Amherst, Massachusetts, gives an annual lecture on the number 17.

The first prime equal to the sum of two consecutive composite numbers. [Beedassy]

The smallest emirp for which both associated primes are the lesser in a twin prime pair. [Capelle]

The Parthenon is 17 columns long.

Only 17 original copies of the *Magna Carta* are known to survive. Texas billionaire and ex-presidential candidate Ross Perot once owned a copy.

There are 17 distinct sets of regular polygons that can be packed around a point. [Astle]

In 1796, Gauss discovered that the regular 17-gon was constructible with compass and straightedge. He was so proud of his discovery that he requested it be carved on his tombstone (like the sphere inscribed in a cylinder on Archimedes'). But it is not there, though neither is his brain, which is in a jar at the University of Göttingen.

17 is the only positive prime Genocchi number. Genocchi numbers are given by the following generating function.

$$\frac{2t}{e^t + 1} = \sum_{n=1}^{\infty} G_n \frac{t^n}{n!}$$

[Terr]

19

The smallest prime that is equal to the product of its digits plus the sum of its digits: $19 = (1 \cdot 9) + (1 + 9)$. [Losnak]

Egyptian biochemist Rashad Khalifa (1935–1990) claimed that he discovered an intricate numerical pattern in the text of the *Qur'an* involving the number 19. Sura 74:30 reads: "Over it are nineteen."

Consider a chess endgame where a King plus Opposite-Colored Bishops (i.e., two Bishops each residing on opposite-colored squares) versus a King. The checkmate requires, at most, 19 moves (if the side with the bishops move first).

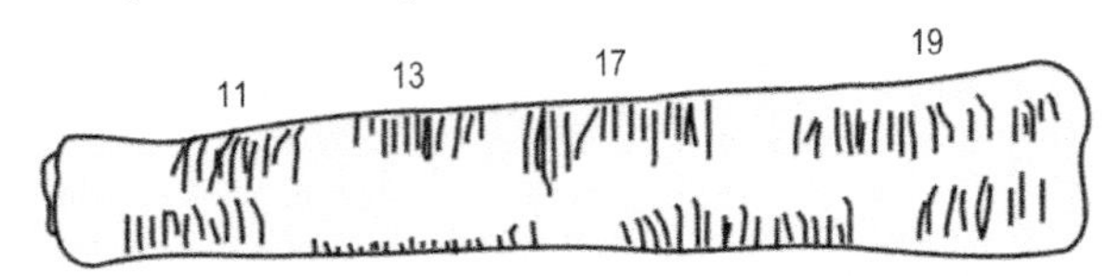

Figure 11. One Side of the Ishango Bone

The largest prime on the Ishango bone (Figure 11). This tool (made from the fibula of a baboon) was found on the shore of Africa's Lake Edward and is believed to be at least 20000 years old. The numbers on one of its columns form a prime quadruple (see k-tuple). It is now located on the 19th floor of the Royal Belgium Institute of Natural Sciences in Brussels.

For decades, mathematicians the world over would open their doors to find the homeless Paul Erdős (1913–1996) announcing, "My brain is open!" It is rumored that for 19 hours a day, seven days a week, stimulated by coffee, and later by amphetamines, he worked on mathematics. One of his greatest achievements was the discovery of an elementary proof for the **prime number theorem**.

The first prime repfigit number. A repfigit (*rep*etitive *fi*bonacci-like di*git*) number is an n-digit integer N with the following property: if a Fibonacci-like sequence (in which each term in the sequence is the sum of the n previous terms) is formed, with the first n terms being the decimal digits of the number N, then N itself occurs as a term in the sequence. For example, if the digits of 19 start a Fibonacci-like sequence, then 19 appears as a term: 1, 9, 10, 19. These are also known as Keith numbers. [Beedassy]

The Bahá'í calendar, established in the middle of the 19th century, is based on cycles of 19 years. Years are composed of 19 months of 19 days each. [Dybwad]

19 is the smallest prime p such that p and p^2 have the same sum of digits.

"The Sun" is numbered with 19 in Tarot cards.

Blaise Pascal deduced 19 theorems related to his famous triangle.

A street in Rome named St. John's Lane is only 19 inches wide.

19 European nations endorsed the first international ban on human cloning.

$2^1 + 3^2 + 5^3 + 7^4 + 11^5 + 13^6 + 17^7 + 19^8$ is prime.

Product 19 is a multi-vitamin and mineral cereal of toasted corn, oats, wheat, and rice.

The Rhind papyrus contained a problem to find x so that x plus one seventh of x will equal 19.

$19 = 8 + 2 + 8 + 1$, and its reversal equals the square root of 8281.

Pont du Gard aqueduct in southern France was built in 19 B.C. [Feather]

In golf, the clubhouse bar is referred to as the 19th Hole.

A gene on Chromosome 19 has been linked to Alzheimer's disease.

In the game of Go, two players alternate in placing black and white stones on a large (19-by-19 line) ruled board, with the aim of surrounding territory.

The smallest number of neutrons for which there is no stable isotope. [Hartley]

Waring conjectured in 1770 that every positive integer can be expressed as a sum of at most 19 biquadrates (fourth powers). This was later proven by Balasubramanian, Deshouillers, and Dress.

Scores of people once died from the Anthrax bacterium following an accident at "Compound 19," a Soviet military facility in the city of Sverdlovsk (now called Yekaterinburg).

19 times its reversal is the famous Hardy-Ramanujan number. Why famous? Because it illustrated Ramanujan's amazing familiarity with the numbers. Hardy wrote:

> "I remember once going to see him when he was ill at Putney. I had ridden in taxi cab number 1729 and remarked that the number seemed to me rather a dull one, and that I hoped it was not an unfavorable omen. 'No,' he replied, 'it is a very interesting number; it is the smallest number expressible as the sum of two cubes in two different ways.'"

Osama bin Laden has 19 brothers. [NBC News]

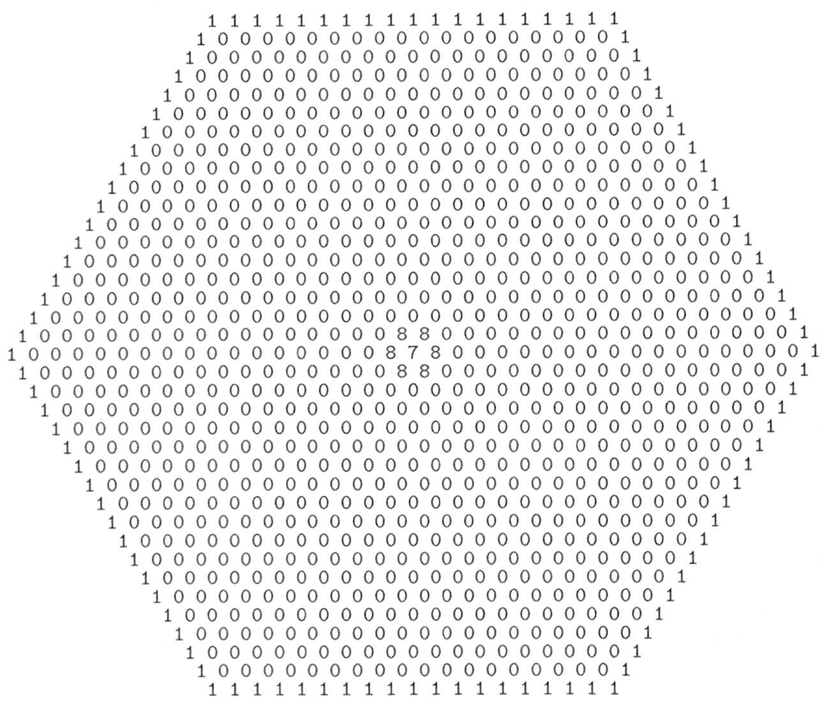

Figure 12. The Smallest Titanic Hexagonal-Congruent Prime

The smallest titanic hexagonal-**congruent prime** has 19 ones on each of its six sides (see Figure 12).

Steely Dan had a hit song called "Hey Nineteen."

The main building of the North Dakota State Capitol is a 19-story Art Deco skyscraper. It is the tallest building in North Dakota in terms of stories. [Bergane]

The only prime that is equal to the difference of two prime cubes. [Gupta]

The decimal expansion of 19^{19} begins with the digits 19. This is not true for p^p, where p is any other prime less than 19,000,000,019.

"19" is the term used to describe a worthless hand in the game of cribbage. [Cary]

The Vice-President of the United States rates a 19-gun salute. [Dobb]

Sigmund Freud was 19 years older than Carl Jung.

19 is the $|1-9|$th prime number.

Professor Barabási and his team have found that the World Wide Web on average has 19 clicks of separation between webpages. [McAlee]

The Fractran algorithm (John Horton Conway's prime-producing machine) is an interesting but terribly inefficient way to generate prime numbers. Start with 2, and then repeatedly multiply the current number at a given stage by the first fraction in the list below that gives an integer value (2, 15, 825, 725, 1925, ...)

$$\frac{17}{91}, \frac{78}{85}, \frac{19}{51}, \frac{23}{38}, \frac{29}{33}, \frac{77}{29}, \frac{95}{23}, \frac{77}{19}, \frac{1}{17}, \frac{11}{13}, \frac{13}{11}, \frac{15}{2}, \frac{1}{7}, \frac{55}{1}$$

Conway showed that the powers of two (other than 2 itself) that occur in this sequence are those with prime exponents. It requires 19 steps to compute the first prime, 2.

The recurring decimal cycles for $\frac{1}{19}$ to $\frac{19-1}{19}$ form a true magic square.

The following are primes: 19, 109, 1009, 10009. No other digit can replace the 9 and yield four primes. [This is the first entry listed as a "Prime Curio" in the book *NUMBERS: Fun & Facts* by J. Newton Friend; Charles Scribner's Sons, New York (1954); Library of Congress Catalog No. 54-8690, p. 45.]

D. H. Lehmer's electromechanical number sieve used 19 bicycle chains to factor numbers. [Bell]

An unistable polyhedron is stable on only one face. The simplest such polyhedron known requires 19 faces. Whether unistability is possible with fewer faces is an unsolved problem. [McCranie]

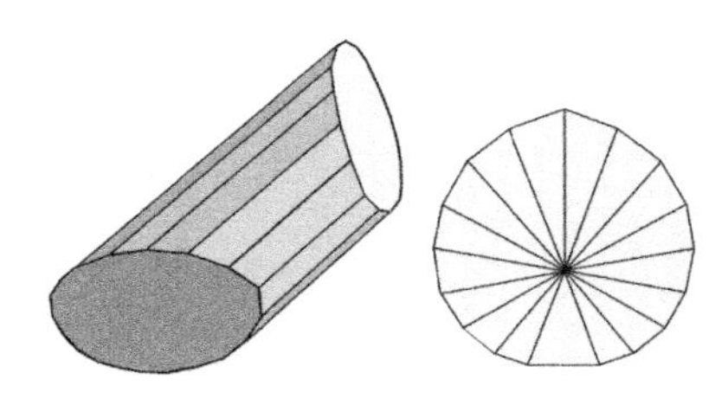

The smallest **invertible prime**.

The sum of the first powers of 9 and 10 and the difference between the second powers of 9 and 10. [Gardner]

$19^0 + 19^1 + \ldots + 19^{18}$ (19 terms) is prime.

23

The smallest prime whose reversal is a power: $32 = 2^5$. [Trigg]

"Hilbert's problems" are a list of 23 problems put forth by David Hilbert in the year 1900. Note the number of words in the following quote attributed to him: "If I were to awaken after having slept for a thousand years, my first question would be: has the Riemann hypothesis been proven?" [Beedassy]

The smallest isolated prime, i.e., not an element of a set of twin primes. [Francis]

Every positive integer is the sum of at most nine cubes. Only 23 and 239 require all nine! Waring's problem (1770) asks: if given k, is there some number $g(k)$ for which every positive integer can be written as the sum $g(k)$ (or fewer) kth powers? For example, $g(3) = 9$ and $g(4) = 19$. In 1909, Hilbert proved the answer was yes.

Homo sapiens have 23 pairs of chromosomes.

23 is the smallest prime for which the sum of the squares of its digits is also an odd prime. [Trotter]

The suicidal bomber has seat number 23 in the film *Airport* (1970).

The smallest prime number that is not the sum of two Ulam numbers. The (standard) Ulam numbers start with 1 and 2; then the subsequent terms are the smallest numbers that can be expressed in just one way as the sum of two distinct earlier terms. [Guy]

The "23 Enigma" is a belief that the number 23 is of particular or unusual significance. (Look up the term *apophenia* when you get a chance.)

The smallest prime that is both preceded and followed by a prime number of successive composites. [Beedassy]

There are 23 discs in the human spine. [McCranie]

W is the 23rd letter of the modern English alphabet. Have you ever noticed that it has 2 points down and 3 points up?

There are 23 definitions in Book I of Euclid's *Elements*.

Euclid (c.300 B.C.)

$\pi(23) = 3^2$.

$23 = 5 + 7 + 11$. Do you see the first five consecutive primes? [La Haye]

23 is the smallest prime number with consecutive digits. [Capelle]

At the height of his career, Professor John F. Nash, Jr., interrupted a lecture to announce that a photo of Pope John XXIII on the cover of LIFE magazine was actually himself (Nash) in disguise and that he knew this because 23 was his favorite prime number. [Hageman]

The largest integer that cannot be expressed as the sum of two squareful numbers. A number is squareful (or non-squarefree) if it contains at least one square in its prime factorization. [Rupinski]

"Twenty-three, skidoo!" is an American catch phrase of unknown origin.

No one really knows exactly when the great English poet and playwright William Shakespeare (1564–1616) was born, but, traditionally St. George's Day (April 23) has been his accepted birthday. However, we do know that he died on this date. [Dowdy]

"The Immortal Game" of chess, played in London on June 21, 1851, between Adolf Anderssen and Lionel Kieseritzky, ended after Anderssen sacrificed a bishop, two rooks, and his queen to deliver checkmate in 23 moves. On the right we show a record of this in algebraic notation.

Anderssen–Kieseritzky *King's Gambit*
1. e4 e5
2. f4 exf4
3. Bc4 Qh4+
4. Kf1 b5
5. Bxb5 Nf6
6. Nf3 Qh6
7. d3 Nh5
8. Nh4 Qg5
9. Nf5 c6
10. g4 Nf6
11. Rg1 cxb5
12. h4 Qg6
13. h5 Qg5
14. Qf3 Ng8
15. Bxf4 Qf6
16. Nc3 Bc5
17. Nd5 Qxb2
18. Bd6 Bxg1
19. e5 Qxa1+
20. Ke2 Na6
21. Nxg7+ Kd8
22. Qf6+ Nxf6
23. Be7 mate

In a room of just 23 people, a greater than 50% chance exists that two of the people will share a common birthday.

Archbishop Ussher (1581–1656), Primate of All Ireland, argued that the world was created on October 23, 4004 B.C. [Delval]

In the Charles Dickens novel *A Tale of Two Cities*, each beheading was counted down. Sidney Carton, the insolent, indifferent, and alcoholic attorney, was guillotine victim number 23. [Nunes]

Julius Caesar was stabbed 23 times when he was assassinated.

The pseudoscientific "biorhythm theory" claims everyone has a 23-day physical cycle that influences their general physical condition.

23 is 2 (mod 3) and 3 (mod 2). [Losnak]

The only prime in a sequence of six numbers that regularly occurred in the first season of the television series *Lost*. [Hartley]

If you were able to fold (doubling each time) a standard sheet of writing paper 23 times, it would become over a mile thick. [Rogowski]

October 23 is Mole Day. It is celebrated among chemists in North America from 6:02 A.M. to 6:02 P.M. on 10/23 each year in honor of Avogadro's number, which is approximately $6.02 \cdot 10^{23}$. [Pritesh]

23 differs from its reversal by 3^2. [Markowitz]

$23 = 1! + (2! + 2!) + (3! + 3! + 3!)$. [Capelle]

29

The smallest prime equal to the sum of three consecutive squares: $2^2 + 3^2 + 4^2$. [Schlesinger]

29 is congruent to 2 (mod 9).

Bobby Fischer was 29 years old at the time of his 1972 World Chess Championship victory in Reykjavík, Iceland. 29 is the largest prime factor of 1972. Was this "Fischer's prime?"

TWENTY NINE can be written out with exactly 29 toothpicks.

According to Gauss, "There have been only three epoch-making mathematicians: Archimedes, Newton, and Eisenstein." Why have so few heard of the latter? Perhaps because Ferdinand Eisenstein died at the age of 29.

Eisenstein (1823–1852)

Floccinaucinihilipilification contains 29 letters and is the longest nontechnical word in the first edition of the Oxford English Dictionary. Note that the letter *e* does not occur.

The first time that a mean prime **gap** (average number of composites between successive primes) is a positive integer occurs at 29. (See Table 11 on page 184.) [Honaker]

29 is the smallest prime of the form $7n + 1$.

The extra day in a leap year is February 29. [Barnhart]

The number of visible notches on the Lebombo bone. This mathematical object was found at Border Cave in the Lebombo mountains of South Africa and is believed to be at least 37000 years old. It resembles the calendar sticks still used today by the Bushmen of Namibia. [Joseph]

The Danish and Norwegian alphabet consists of 29 letters.

Euclid's 29th Proposition (*Elements*, Book I) was the first one to use his parallel postulate. [McCranie]

29 is the maximum number of squares a chess bishop can visit if it is only allowed to visit each square once. (Here, "visit" means that squares passed over in a move are also visited.)

The highest possible score of a single hand in the card game cribbage is 29. [McGowan]

$29 = \sqrt{6! + \frac{6!+6}{6}}$. [Mensa]

The B-29 Superfortress was a long-range heavy bomber aircraft flown by the U.S. Military in World War II and the Korean War. [Haga]

The last prime in the smallest set of five primes in arithmetic progression. (A sequence is an arithmetic progression if each term is the preceding term plus a fixed common difference, e.g., a, $a + d$, $a + 2d$, $a + 3d$, ...)

$2n^2 + 29$ is prime for $n = 0$ to 28. [Legendre]

29 people lost their life when the SS *Edmund Fitzgerald* sank on Lake Superior (17 miles north-northwest of Whitefish Point, Michigan; November 10, 1975). [McCranie]

The legendary Mississippi blues musician Robert Johnson (1911–1938) recorded only 29 songs. [McCranie]

Twentynine Palms (or 29 Palms), California, is located approximately halfway between Los Angeles and Las Vegas and is considered by many to be a prime destination. [Jen]

The Oxyrhynchus papyrus containing a complete diagram (dated A.D. 75–125) from Euclid's *Elements* has a red 29 on it that was presumably written there by B. P. Grenfell or A. S. Hunt of Oxford University. The fragment is now in the storage vaults of the Penn Museum (University of Pennsylvania Museum of Archaeology and Anthropology).

$7^2 + 8^2 + \ldots + 28^2 + 29^2$ is a square. [Capelle]

The middle chapter of the Old Testament is Job 29.

Track 29 is the line on which the display locomotive "Chattanooga Choo Choo" now rests in Chattanooga, Tennessee. [McCranie]

Mrs. Prime reads her recantation in Chapter 29 of the 19th century classic *Rachel Ray* by British novelist Anthony Trollope.

There are 29 suras of the *Qur'an* that are prefixed with certain letters of the Arabic alphabet (or Qur'anic Initials). [Yuksel]

The IBM 29 card punch was first sold in 1964. It continued to be sold for two full decades.

31

There are only 31 numbers that cannot be expressed as the sum of distinct squares.

The number of letters (in English) required to write the word names of the first six primes is the sixth prime reversed, i.e., 31. [Trotter]

Buddha preached on the 31 levels of existence in our universe. [Jen]

A computer analysis by Ströhlein and Zagler shows that the winning process in a Queen versus Rook endgame should take at most 31 moves.

π^3 is close to 31. [Luhn]

The Mir Space Station docked with 31 spacecraft during its history.

Portal 31 is Kentucky's first exhibition coal mine.

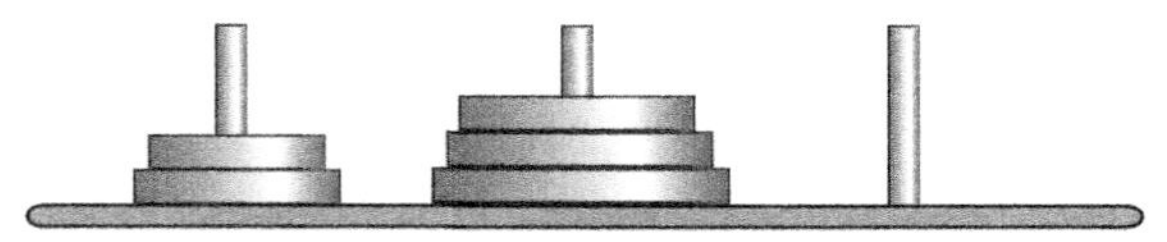

Figure 13. The Tower of Hanoi Puzzle

The minimal number of moves required in a Tower of Hanoi puzzle containing 5 discs (Figure 13). The general solution to this puzzle requires a **Mersenne number** (i.e., $2^n - 1$) of moves, where n is the number of discs.

The smallest multidigit prime whose reciprocal has an odd period for its decimal expansion.

Mersenne prime $M(31)$ could easily be confused with our nearest large neighbor galaxy Andromeda, i.e., M31 (or Messier 31).

$31 = 2^2 + 3^3$. [Kulsha]

The smallest prime such that replacing each digit d with d copies of the digit d produces a different prime (3331).

There are 31 milligrams of cholesterol in a tablespoon of butter.

The only known Mersenne emirp.

The speed limit in downtown Trenton, a small city in northwestern Tennessee, is 31 miles per hour.

31 letters of the Russian alphabet are pronounceable.

$3 + 5 + 7 + 11 + \ldots + 89 = 31^2$, and the sum of the first 31 odd primes is a prime square.

The sum of digits of the 31st Fibonacci number is 31. [Gupta]

The big "31" sign made its debut at all Baskin-Robbins stores in 1953, offering customers a different ice cream for every day of the month. Note that 31 is the largest prime factor of 1953. [Coneglan]

The first U.S. space satellite (Explorer-I) weighed just less than 31 pounds and was launched on Jan 31, 1958. The high-power transmitter worked for 31 days. [McCranie]

31 and the 31st prime are both Mersenne primes. [Wu]

The smallest prime that is a **generalized repunit** in three different bases. [Rupinski]

The smallest prime that can be represented as the sum of two triangular numbers in two different ways (21 + 10 and 28 + 3). [Gupta]

Figure 14. Moser's Circle Problem

Moser's circle problem asks to determine the most pieces into which a circle is divided if n points on its circumference are joined by chords with no three internally concurrent. The first few values are 1, 2, 4,

8, and 16 (see Figure 14). What do you think the next term will be? [Post]

Jackson Pollock's *One: Number 31, 1950* occupies an entire wall by itself at the Museum of Modern Art, New York City. The painting is oil and enamel on unprimed canvas. (Studies have shown that some of Pollock's works display the properties of fractals.)

The number of regular polygons with an odd number of sides that can be constructed with compass and straightedge.

There are 31 pairs of spinal nerves in the human body. [Beedassy]

After removing two opposing corner squares on a chessboard it is impossible to arrange 31 dominoes in such a way that they cover all the remaining squares.

31 squared when turned upside down is a perfect square. [Friend]

961 196

In the *Star Trek* episode "Terra Prime," Reed's Section 31 contact provides information to help the *Enterprise* crew infiltrate a Martian colony.

Ben Johnston is known for composing music based on a flexible tuning system that derives pitches from as high as the 31-limit. The "prime limit" of an interval or chord in just intonation is the largest prime number in its factorization.

31 is the largest integer n such that the first n digits of π after the decimal point are all nonzero. [Keith]

37

Mitochondrial DNA commonly found in most animals contains 37 genes. [Magliery]

The German physician, pioneer psychiatrist, and medical professor Carl Reinhold August Wunderlich (1815–1877) is best-known for his observations and conclusion that the mean healthy human body temperature is 37 °C.

$2^3 + 5^7 + 11^{13} + 17^{19} + 23^{29} + 31^{37}$ is prime.

$37 \cdot 03 = 111$	$37 \cdot 12 = 444$	$37 \cdot 21 = 777$
$37 \cdot 06 = 222$	$37 \cdot 15 = 555$	$37 \cdot 24 = 888$
$37 \cdot 09 = 333$	$37 \cdot 18 = 666$	$37 \cdot 27 = 999$

Figure 15. Products of 37

Euler (1707-1783)

The largest prime number found in Euler's list of 65 "numeri idonei" (37 of the 65 are squarefree). The integer m is an idoneal number (also called a convenient number) if every odd number $n > 1$ that can be written uniquely in the form $x^2 + my^2$ with $x, y \geq 0$ and $\gcd(x, my) = 1$, must be a prime or a prime power. It is believed, but not yet proven, that Euler's list (Table 8 on page 154) is complete. [Younce]

William Shakespeare is thought to have written 37 plays.

All three-digit repdigits are divisible by 37 (Figure 15).

The French version of solitaire uses a board with holes for 37 pegs.

In 1989, the "Amdahl Six" team found a prime larger than Slowinski's previous record prime by just 37 digits. It was the only non-Mersenne prime to be the largest known prime since 1952.

Pliny the Elder (A.D. 23–79) wrote many historical and technical works, but only his 37-volume *Historia Naturalis* (Natural History) has survived. [Hadas]

Ramanujan had 37 papers published in peer-reviewed mathematical journals. [Croll]

37 and the 37th Mersenne prime exponent are emirps. [Wu]

In the movie *The Mothman Prophecies*, Connie Parker has a prophetic dream in which she hears whispered to her "wake up, number 37." [La Haye]

In 1953, baseball great Ted Williams played in only 37 games and had 37 hits. [Rachlin]

The late 1960's serial killer who referred to himself as "Zodiac" claimed to have taken 37 lives in the San Francisco Bay area.

"37 Heaven" is an online collection of 37 factoids.

$$\begin{aligned}(k+2)\{1 &- [wz+h+j-q]^2 - [(gk+2g+k+1)(h+j)+h-z]^2 \\ &- [2n+p+q+z-e]^2 - [16(k+1)^3(k+2)(n+1)^2+1-f^2]^2 \\ &- [e^3(e+2)(a+1)^2+1-o^2]^2 - [(a^2-1)y^2+1-x^2]^2 \\ &- [16r^2y^4(a^2-1)+1-u^2]^2 - [n+l+v-y]^2 \\ &- [((a+u^2(u^2-a))^2-1)(n+4dy)^2+1-(x+cu)^2]^2 \\ &- [(a^2-1)l^2+1-m^2]^2 - [ai+k+1-l-i]^2 \\ &- [p+l(a-n-1)+b(2an+2a-n^2-2n-2)-m]^2 \\ &- [q+y(a-p-1)+s(2ap+2a-p^2-2p-2)-x]^2 \\ &- [z+pl(a-p)+t(2ap-p^2-1)-pm]^2\}\end{aligned}$$

Figure 16. All positive values are prime, but can you find one?

The only prime with period length three: $\frac{1}{37} = 0.\overline{027}$. A prime which has a period length that it shares with no other prime is called a **unique prime**.

The degree of the first polynomial discovered whose set of positive values is the set of primes as the variables range over the natural numbers. Matijasevič showed this was possible in 1971, and Jones, Sato, Wada, and Wiens provided an example of such a polynomial with 26 variables (and degree 25) in 1976. It can conveniently be written down using the 26 letters of the English alphabet (Figure 16).

The Carthaginian general Hannibal began his long march across the Pyrenees with 37 war elephants. [Polybius]

The sum of the first 37 primes is a Fibonacci number.

$37 = 33 + 3 + \frac{3}{3}$.

The sum of the first five consecutive composite numbers (4, 6, 8, 9, and 10). [Natch]

In the book *A Painted House* by John Grisham, the character Luke observed that his grandfather always drove his truck at exactly 37 miles per hour. Did it run most efficiently at this speed?

$4^n + 37$ yields primes for $n = 1, 2, 3, 4, 5, 6,$ and 7. [Trotter]

The smallest **irregular prime**. [Russo]

European Roulette is played using a wheel containing 37 numbered slots (1 to 36, plus a 0).

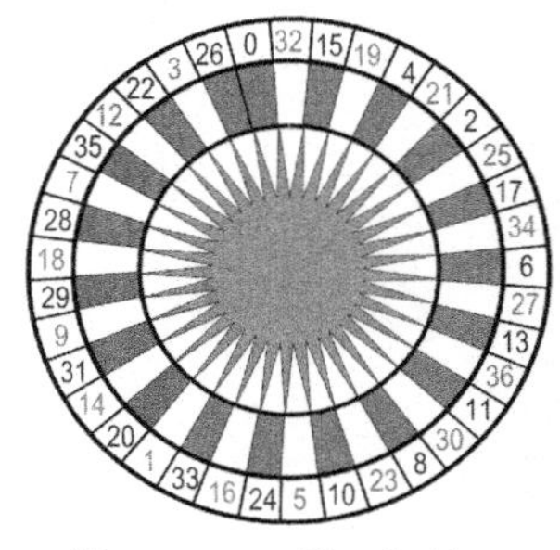

European Roulette

Physicist and Nobel laureate Richard Feynman (1918–1988) wrote only 37 research papers in his career.

In the movie *Office Space*, the waiter is wearing 37 pieces of flair. [Patterson]

Blue Moons occur 37 times every century on average. [Cooley]

All positive integers can be written as the sum of not more than 37 fifth powers (Chen Jingrun, 1964).

$\pi(37) = 12$ and $\pi(73) = 21$. [Pe]

"37signals" is a privately held web design and web application company based in Chicago, Illinois. The company is named for 37 radio telescope signals identified as potential messages from extraterrestrial intelligence.

The following 37-word quotation is ascribed to the great Swiss mathematician Leonhard Euler (1707–1783): "Mathematicians have tried in vain to this day to discover some order in the sequence of prime numbers, and we have reason to believe that it is a mystery into which the human mind will never penetrate."

The last odd Roman numeral alphabetically is XXXVII (37). [Brahinsky]

The prime $p = 37$, and its reversal $q = 73$, are the only known emirp pair such that $p! + 1$ and $q! + 1$ are both primes. [Punches]

41

Prince Judah, the title character in the epic film *Ben-Hur*, is reduced to oarsman number 41, a nameless slave on a Roman battleship.

The number of whacks accused murderess Lizzie Borden allegedly gave her father on August 4, 1892, in Fall River, Massachusetts. She

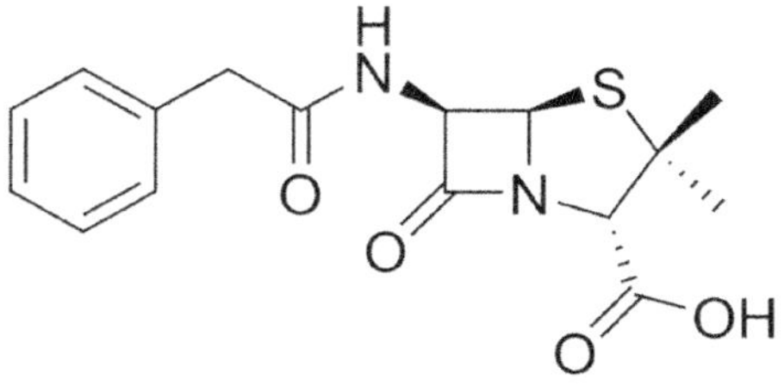

Figure 17. Chemical Structure of Penicillin G's 41 Atoms

was acquitted of the murders of her father and stepmother, and the house where the murders occurred is now a bed and breakfast.

$41 + 14!$ is prime.

A molecule of benzylpenicillin (commonly known as penicillin G) has 41 atoms (Figure 17). [Scott]

A prime p is a "lucky number of Euler" if the values of $n^2 + n + p$ are all prime for n from 0 to $p - 2$. There are only six of these and the largest is 41. The others are 2, 3, 5, 11, and 17. [Le Lionnais]

$41! + 1$ is prime.

The sum of digits and period length of the reciprocal of 41 are equal.

The smallest prime such that the sum of factorials of its distinct digits is a square number ($4! + 1! = 5^2$).

The 41st heptagonal number ($r(5r - 3)/2$, for $r = 41$) is the concatenation of 41 with itself. [Heleen]

Robert Goddard successfully launched the world's first liquid-fueled rocket on March 16, 1926. It reached an altitude of 41 feet. [Collins]

41 is the largest known prime formed by the sum of the first Mersenne primes in logical order ($3 + 7 + 31$). [Luhn]

For every pair of positive numbers that sum to 41, the square of the first one plus the second one is prime. [Chubio]

Sum 41 is a video-obsessed pop punk band from Canada.

The Sun's wobble speed is about 41 feet per second. [Marcy]

The Old Homestead Steakhouse in New York City claims their theory is simple: "Nothing but *prime*, dry aged and grilled to perfection." Their domestically-raised, hand massaged Kobe beef and world famous $41 hamburger are trademarked. [Nussbaum]

Symphony number 41 (*Jupiter*) was the last symphony that Wolfgang Amadeus Mozart wrote. [Obeidin]

There is a 41 Gun Royal Salute in London for a Royal Birthday, a Royal Birth, the State Opening of Parliament, and the arrival of a foreign Head of State or Commonwealth Prime Minister.

Captain Cook lost 41 of his crew to scurvy on his first voyage to the South Pacific in 1768.

A prime number of length 41 can be constructed by concatenating the exponents of the first consecutive Mersenne primes, separated by 0's. (Hint: the number of exponents used is the reversal of 41.) [Hartley]

The number of votes required to sustain a filibuster in the United States Senate. [McCranie]

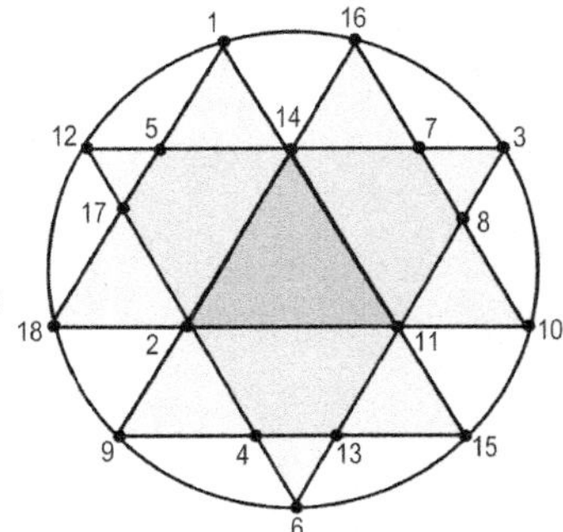

The Durga Yantra is a symbolic image used to represent the goddess Durga (the first female divinity according to Hindu scriptures). It is possible to place the numbers from 1 to 18 at the intersections of this image so that the sum along each line segment is 41. [Keith]

43

London was founded as the Roman town of Londinium in A.D. 43.

The first 43 digits of 43! form a prime number.

There exist three different two-digit prime numbers such that the average of all three is 43 and the average of any two is also prime. [Sole and Marshall]

43 women survived at the Battle of Dannoura, yet the entire Heike battle fleet was destroyed. This occurred at the Japanese Inland Sea in April of 1185.

The British mathematician and logician Augustus De Morgan was always interested in odd numerical facts and once noted that he had the distinction of being x years old in the year x^2. He was 43 in 1849. [Poo Sung]

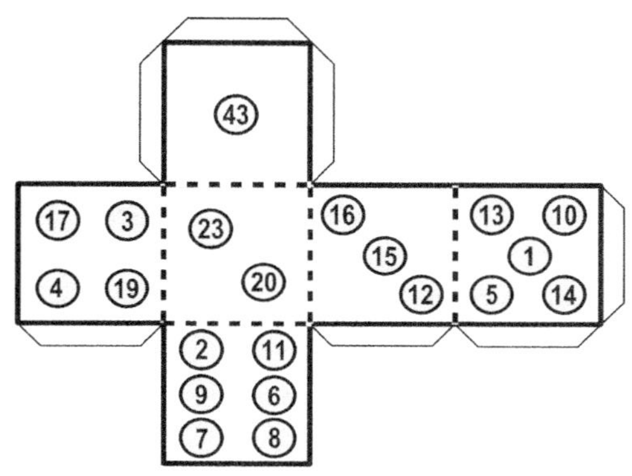

Figure 18. Pattern for a Honaker Die

The smallest possible magic sum if you number the pips (dots) on an ordinary six-sided die with distinct positive integers such that the numbers on each face add up to the same magic constant (Figure 18). Can you find different sets of numbers for this puzzle that add to 43?

The German-born Jewish mathematician Emmy Noether (1882–1935) published 43 research papers. Albert Einstein eulogized her in a letter that appeared in the New York Times on May 5, 1935, where he is quoted as saying, "In the judgment of the most competent living mathematicians, Fräulein Noether was the most significant creative mathematical genius thus far produced since the higher education of women began." [Wheeler]

Licor 43, also known as "Cuarenta Y Tres" (meaning 43 in Spanish), is a liqueur that has been made in Spain for over 1800 years, dating back to the time of the Carthaginians. It is a bright yellow color and derives its name from the fact that it is supposedly a mixture of 43 different ingredients.

The $(4 \cdot 3)$th lucky number. [Post]

A frown uses 43 muscles. :-(

George Parker Bidder (1806–1878), an English calculating prodigy, could instantly recite a 43-digit number after it had been read to him backwards.

The Antikythera mechanism (described as the first known mechanical computer) was found in less than 43 meters of water. [Russell]

The NASCAR Winston Cup Series races begin with 43 cars and drivers. [Dobb]

Leonard Owen Johnson Associates reported in 2001 that tobacco smoke contains at least 43 chemicals that are known carcinogens.

43 appears in the title of the historical British comedy show *Hancock's Forty-Three Minutes.*

$43^7 = 271818611107 = (2+7+1+8+1+8+6+1+1+1+0+7)^7$. [Madachy]

Barbie and Ken, the most popular fashion dolls in the world, broke up after 43 years. This was probably due to Ken's reluctance to get married. [Cross]

It was 43 seconds after "Little Boy" was released from *Enola Gay* that the mechanisms aboard the first nuclear weapon gave the signal to detonate over Hiroshima on August 6, 1945.

43
4483
444883
44448883
4444488883
are 5 primes

Technetium (atomic number 43) is a crystalline metallic element and the first to be produced artificially.

The 14th prime number. Note that the previous prime (41) is the reversal of 14.

What is the minimum number of guests that must be invited to a party so that there are either five mutual acquaintances, or five that are mutual strangers? At least 43, but the exact number is not yet known. The solutions to this type of problem are known as Ramsey numbers, named for British mathematician Frank P. Ramsey (1903–1930).

The smallest prime that is not a **Chen prime**. [Lopez]

Ostriches can sprint in short bursts up to 43 miles per hour.

The blog "Interstate Forty-Three" is dedicated to the number 43.

Fields medalist Alexander Grothendieck (1928–) is highly regarded as one of the most influential mathematicians of modern times. The subject of many stories and rumors, he withdrew from mathematics just before turning 43 years of age and now avoids virtually all human contact.

The number 'forty-three' is the smallest prime not mentioned in the KJV Bible text.

47

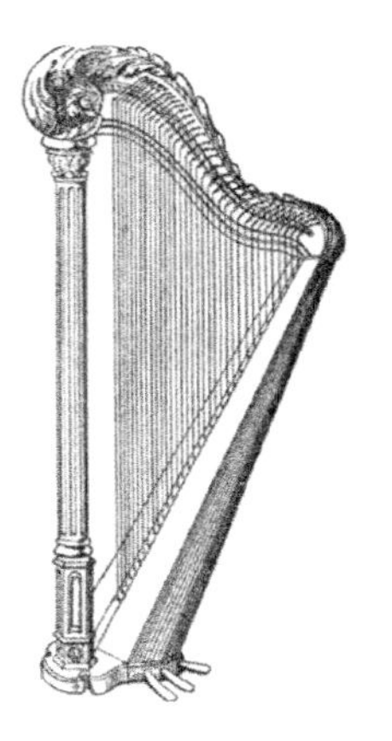

The modern concert harp typically has 47 strings. [Kamath]

$47 + 2$ equals the reversal of $47 \cdot 2$.

Mathematician Kevin Hare has proven that any odd perfect number must have at least 47 prime factors, including repetitions.

There is an abnormally high use of the prime number 47 in episodes of *Star Trek*. For example, in the *Star Trek: Voyager* episode "Infinite Regress," Naomi Wildman reveals that there are 47 sub-orders of the Prime Directive.

The house band for New York City called Black 47 resides on 47th Street.

The AK-47 is the most widespread weapon in the world. [Kalashnikov]

John Major was 47 years old when he became British Prime Minister in 1990.

The MK-47 was the first and last Soviet calculator with magnetic cards. [Frolov]

The KJV Bible was translated by 47 scholars. [English]

There are 47 Pythagorean propositions in Euclid's *Elements*. [Dobb]

Isaac Asimov's Book of Facts mentions the "fact" that mosquitoes have 47 teeth. [Patterson]

Mary Mallon, better known as "Typhoid Mary," was a food service worker who infected 47 people with the bacterium *Salmonella enterica typhi.* She was the first "healthy carrier" of typhoid fever in the United States.

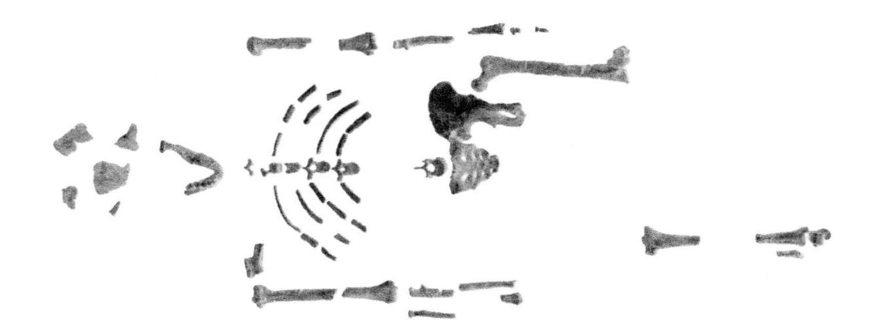

Figure 19. The 3.2 Million-Year-Old Bones of "Lucy"

The best-known fossil of *Australopithecus afarensis* was named "Lucy" (after the Beatles' song "Lucy in the Sky with Diamonds"). 47 of her bones were unearthed in Ethiopia in '74 (Figure 19).

According to Chinese legend, a 16th century official named Wan Hu attempted a flight to the Moon by using two kites fastened to a sedan chair on which he had strapped 47 black powder rockets. 47 servants simultaneously lit the fuses and Wan Hu disappeared in a burst of flame and smoke. A crater on the far side of the Moon (at 9.8°S, 138.8°W) has been named in his honor.

Wan Hu's "Flight"

The 47 Society is a humorous society at Pomona College in California that propagates the belief that the number 47 is the quintessential random number. Campus lore suggests that Pomona math professor Donald Bentley produced a convincing mathematical proof that 47 was equal to all other integers. (See `www.47.net/47society/`.) [Schuler]

In the Pixar Film *Monsters Inc.*, the scream factory had enjoyed 47 accident-free days at the start of the movie. [Hartley]

In the play *The Five Hysterical Girls Theorem* by Rinne Groff, renowned number theorist Moses Vazsonyi fears that he has lost his edge in the intellectually grueling world of prime **number theory** at the age of 47.

There are 47 occurrences of 47 in the first thousand prime numbers. [Faust]

Window washer Alcides Moreno survived a 47-story fall from a New York City skyscraper.

The Hadwiger problem sought the largest number of subcubes (not necessarily different) into which a cube *cannot* be divided by plane cuts. In 1947, it was shown the answer was 47 or 54. In 1977, the dissection in Figure 20 was discovered, proving the answer was 47. [Post]

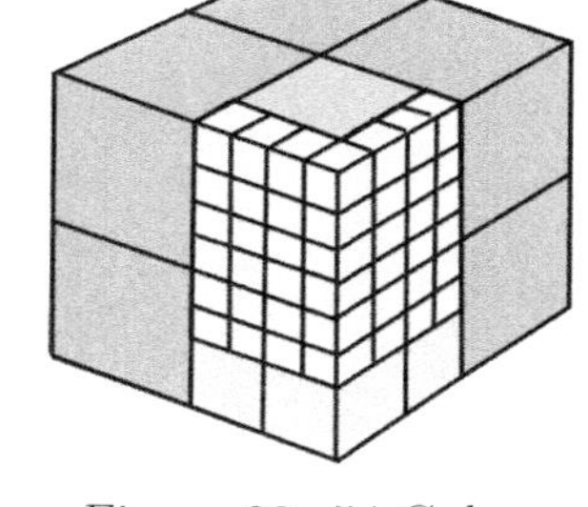

Figure 20. 54 Cubes

Can you find the 4 consecutive Fibonacci numbers whose product equals the product of the first 7 prime numbers? Hint: computing the sum of the four consecutive Fibonacci numbers is as easy as 1-2-3.

Henri Emile Benoit Matisse's painting "Le Bateau" hung for 47 days in the Museum of Modern Art, New York City, between October 18 and December 4, 1961, and not a single person noticed it was upside down.

53

The prayer of Ave Maria is repeated 53 times in the recitation of the rosary. [Desrosiers]

The chance that no pair of 53 people in a room have the same birthday is approximately $\frac{1}{53}$. [ApSimon]

The website address for Fifth Third Bank is `www.53.com`.

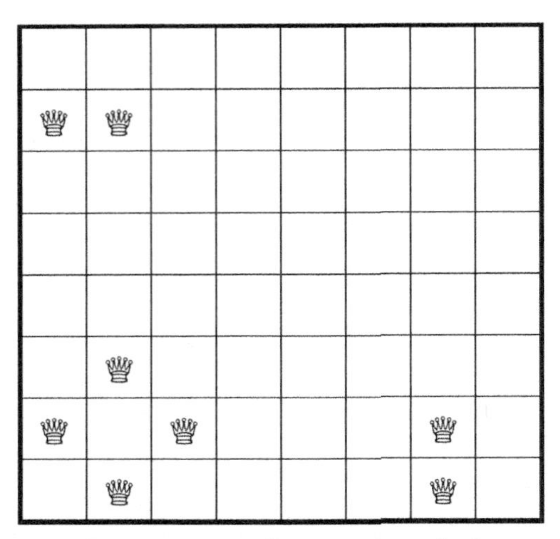

Figure 21. 53 Squares Attacked

It is possible to place eight queens on a standard chessboard in which only 53 squares are under attack (Figure 21). Is this the minimum solution? [Madachy]

$\pi(53) = 5^2 - 3^2$.

The number painted on the Volkswagen movie star, "Herbie the Love Bug."

The reversal of 53 equals the sum of digits of 53^3.

$\pi(3^5) = 53$. [Firoozbakht]

53 is thought to be the smallest prime that is not the sum or difference between powers of the first two prime numbers.

The smallest multidigit balanced prime. [Russo]

ATM (Asynchronous Transfer Mode) is a high-speed network protocol composed of 53 byte "cells."

53 in decimal is 35 in hexadecimal. [Nigrine]

The United States allows the use of 53-foot trailers (see Figure 22) behind a semi-tractor (called a "prime mover" in Australia).

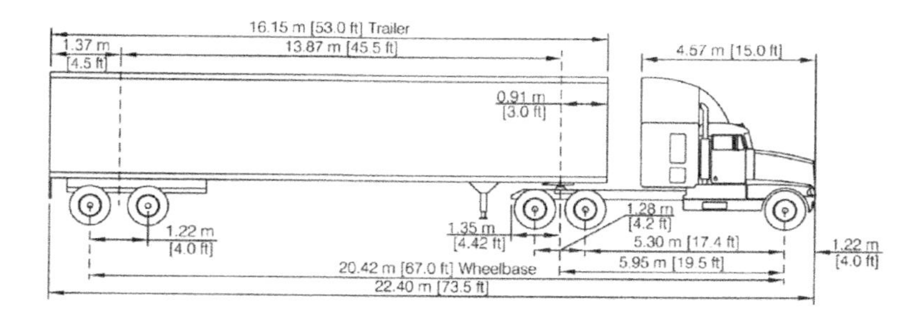

Figure 22. U.S. Federal "Green Book" WB-20 (WB-67): 53' Trailer

59

The smallest "Cypher prime" (pages 235–236) in the KJV Bible. It can be deciphered in the shortest verse (John 11:35). [Bulmer]

The least irregular prime of the form $4n + 3$.

Annika Sorenstam was the first woman golfer to shoot 59 in LPGA competition. [Sturgill]

The first 59 digits of 58^{57} form a prime. [Kulsha]

There are exactly 59 stellations of the regular icosahedron. To the right we present just one example.

The center prime number in a 3-by-3 prime magic square that has the smallest possible magic constant. A discovery attributed to Rudolf Ondrejka.

The digits 1 through 59 in the Babylonian sexagesimal numeral system were not distinct symbols. [Heeren]

Thomas Alva Edison's first Electric Power Plant in New York City supplied 59 customers within a one square mile area. [McAlee]

59 inches equals 1 yard, 1 foot, and 11 inches. [McAlee]

The Bell XP-59 Airacomet was America's first jet-propelled airplane.

The smallest prime in the Pythagorean triangles (59, 1740, 1741) and (61, 1860, 1861), where each right triangle has two prime sides and the smallest pair of primes (59 and 61) are twin primes. [Hess and Palos]

61

The smallest prime whose reversal is a square. [Trigg]

61 divides $67 \cdot 71 + 1$. **Honaker's problem** asks if there are three larger consecutive primes $p < q < r$ such that p divides $qr + 1$.

The number of digits in $M(61)$ is 61 turned upside down.

Louis A. Bloomfield's textbook *How Things Work* examines 61 objects in our everyday world.

You can arrange 61 coins into a hexagonal pattern with one coin in the center.

The 61st book in the KJV Bible has 61 verses. [Desrosiers]

61 codons specify individual amino acids. [Necula]

Ten consecutive distinct digits begin at position 61 after the decimal point of π. [Wu]

One way to convert words to numbers is to let A = 1, B = 2, C = 3, ..., Z = 26; and then compute the sum. Using this "**alphabet code**," the word PRIME is prime.

Roger Maris ended the season with 61 home runs in '61.

The average gestation period of a dog is 61 days.

In 1657, Fermat challenged the mathematicians of Europe and England, "We await these solutions, which, if England or Belgic or Celtic Gaul do not produce, then Narbonese Gaul (Fermat's region) will." Among the challenges was this 500-year-old example from Bhaskara II: $x^2 - 61y^2 = 1$ $(x, y > 0)$. [Balazs]

The letter 'A' occurs 61 times among the U.S. state names. It is the most common letter used in this manner. [Blanchette]

The smallest prime that can be written as the sum of a prime number of primes to prime powers in a prime number of ways: $2^2 + 2^3 + 7^2$, $2^2 + 5^2 + 2^5$, and $3^2 + 5^2 + 3^3$. [Hartley]

The largest known prime of the form $Ack(m, m)$, where Ack is the Ackermann function (which grows faster than an exponential function). $Ack(3, 3) = 61$. [Hartley]

61 is the sum of two consecutive squares: $5^2 + 6^2$. [Schlesinger]

There are 61 abstract groups of order 1464. Note that 61 divides 1464. [Hartley]

Promethium (atomic number 61) may be missing from our solar system, but it has been detected in the spectrum of a variable star in the constellation Andromeda.

61 countries took part in World War II.

In the tiny hamlet of Pinghan, nestled deep among a stand of limestone hills in a remote region of southwestern China, locals honor an old national tradition of buying a coffin at the age of 61.

61 is the $(6 - 1)$th repfigit number: 61, $6 + 1 = 7$, $1 + 7 = 8$, $7 + 8 = 15$, $8 + 15 = 23$, $15 + 23 = 38$, $23 + 38 = 61$. [Mirizzi]

$61! - 60! + 59! - 58! + \ldots + 5! - 4! + 3! - 2! + 1!$ is prime.

The first prime overlooked by Mersenne in his erroneous conjectured list of exponents. [Beedassy]

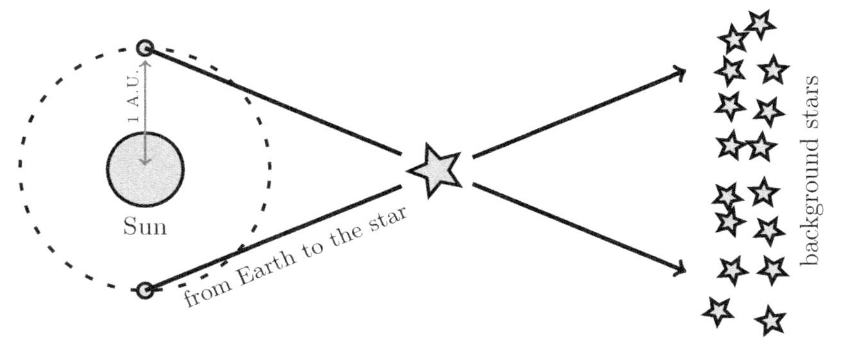

Figure 23. Measuring Distance with Trigonometric Parallax

In 1838, astronomer Friedrich Wilhelm Bessel calculated the large proper motion of 61 Cygni using trigonometric parallax (Figure 23). This was the first star (other than the Sun) to have its distance known from the Earth. [McGown]

It is not known whether the double-Mersenne number $MM(61) = 2^{2^{61}-1} - 1$ is prime or composite. It contains 694,127,911,065,419,642 digits and is currently far too large for the usual Lucas-Lehmer test.

Bob Dylan wrote a song (and album) called "Highway 61 Revisited."

61 (sixty-one) is the largest prime whose name spelled out in English has no repeated letters. [Brahinsky]

67

The largest prime which is not the sum of distinct squares. [Crespi de Valldaura]

Mersenne claimed $2^{67} - 1$ was prime, but it's composite.

*67 ("star 67") is a free phone feature that blocks sending name and phone number on an outgoing individual call. Use *67 + number being dialed (must be done for each call). [Greer]

Using the alphabet code (page 50), the value of 67 in its Roman numeral-based representation (LXVII) is the reversal of 67. [Necula]

The smallest prime p that divides the number of composites less than the $(p + 1)$th prime. [Honaker]

$\pi(6 \cdot 7) = 6 + 7$. [Firoozbakht]

$67 = \sqrt{\frac{9!}{9 \cdot 9} + 9}$. [Poo Sung]

5^{67} starts with the digits 67. [Hartley]

The "Summer of Love" was a phrase given to the summer of '67, describing (personifying) the feeling of being in San Francisco that summer when the hippie movement came to full fruition.

John Napier, the inventor of natural logarithms, died at the age of 67. He is also remembered for a device (known as "Napier's bones") consisting of numerical rods used for calculating products and quotients of numbers. His reputation of being a sorcerer is strengthened by tales of him carrying around a black spider in a small box.

The Escher compound that appears in the "Waterfall" lithograph by M. C. Escher divides the three component cubes into 67 individual cells. [Post]

$6! \cdot 7! = 10!$.

6^7 is the sum of a twin prime pair. [Rivera]

Longitude 67 degrees West passes through the easternmost city in the United States (Eastport, Maine). A prime place to see the sunrise. [Punches]

71

Conway's constant is an algebraic number of degree 71.

$71^2 = 7! + 1!$. [Patterson]

71 divides the sum of primes $2 + 3 + 5 + 7 + 11 + \ldots + 61 + 67 + 71$.

Isaac Newton's *Proposition 71* proves that a homogeneous sphere attracts particles external to the sphere as if all of its mass were concentrated at its center.

Newton (1642–1727)

71 cubed is a concatenation of the first five odd numbers. [Davis]

In 1935, Erdős and Szekeres proved that 71 points (no three on a single line) are required to guarantee there are six that form a convex hexagon, although 17 points are thought to be sufficient. (In 1998, the upper bound was reduced to 37.)

The smallest prime formed from the concatenation of happy numbers in reverse order. [Gupta]

The largest of the supersingular primes, i.e., the set of primes that divide the order of the Monster group (an algebraic construction with $2^{46} \cdot 3^{20} \cdot 5^9 \cdot 7^6 \cdot 11^2 \cdot 13^3 \cdot 17 \cdot 19 \cdot 23 \cdot 29 \cdot 31 \cdot 41 \cdot 47 \cdot 59 \cdot 71$ elements).

The Great Sanhedrin was the supreme religious body in the Land of Israel during the rabbinic period. Tannaitic sources describe it as an assembly of 71 sages.

Clint Eastwood is Inspector 71 in the movie *Dirty Harry* ('71).

The repunit $R_{71} = (10^{71} - 1)/9$ is a semiprime and therefore relatively hard to factor. Note that its smallest factor contains all of the digits, except $7 + 1$.

The smallest prime revrepfigit (*rev*erse *rep*licating *fi*bonacci-like di*git*) number: 7, 1, 8, 9, 17 (see page 26). [Earls]

73

Pi Day (March 14 or 3/14) occurs on the 73rd day of the year on non-leap years. Howard Aiken died on Pi Day at the age of 73. He was the primary engineer behind IBM's Harvard Mark I computer. [Rupinski]

$\pi(73) = 7 \cdot 3$. [Honaker]

Starting with 73, you must repeatedly double and add 1 a palindromic number of times before a prime is reached. [Roonguthai]

The space shuttle Challenger disaster occurred 73 seconds after liftoff.

The Empire State Building has 73 elevators.

73 is the 37th odd number. [Hasler]

Here's a neat trick with its digits: $7^3 = 343$ or $(3 + 4)^3$.

The first difficult prime to form under the rules of the four 4's puzzle. The problem is to find expressions for numbers using exactly four 4's and a finite number of mathematical symbols and operators in common use. One version of the rules allow only addition, subtraction, multiplication, division, square root, exponentiation, factorial, decimal point, parentheses, as well as concatenation. Here are two possible solutions (achieved by altering the rules above):

$$73 = \frac{\left(\frac{4!}{\sqrt{4}}\right)}{4} + 4 = \frac{4! + 4! + \sqrt{.\overline{4}}}{\sqrt{.\overline{4}}}.$$

Why not try to "build" all of the prime numbers below 100? The first few are $\frac{4 \cdot 4}{4+4}$, $\frac{4+4+4}{4}$, $\frac{4 \cdot 4+4}{4}$, and $\frac{44}{4} - 4$.

The minimum number of sixth powers needed to represent every possible integer. [Pillai]

The record number of major league home runs hit in one season is 73. Barry Bonds set it in 2001.

The number formed by the concatenation of odd numbers from 73 down to 1 is prime. [Patterson]

The total number of books in the *Catholic Bible* if the Book of Lamentations is counted separate from the Book of Jeremiah. [Medine]

The number of rows in the Arecibo Message graphic (Figure 24). The Arecibo Message is a radio message that was sent toward a globular star cluster in Hercules from a radio telescope in Puerto Rico on November 16, 1974. The entire transmission lasted a semiprime number of seconds and provided a way for extraterrestrials to arrange a 73-by-23 array into recognizable data about our life and solar system. Can you read it?

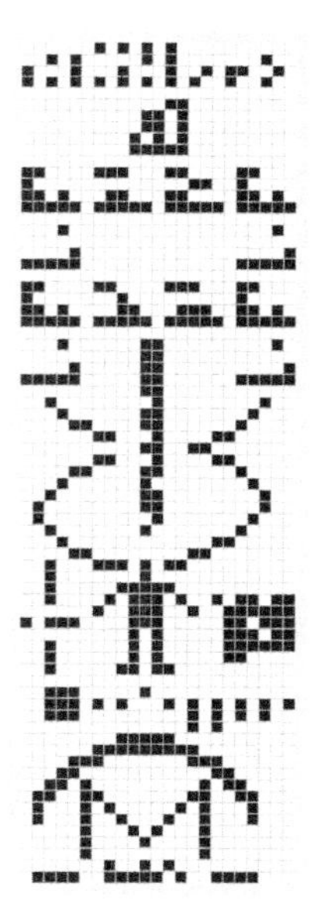

Figure 24.

Thomas Jefferson and Aaron Burr each received 73 electoral votes and tied for the presidency.

Gakuho Abe constructed a pair of magic squares using 73 consecutive prime numbers.

Manjul Bhargava showed that if a quadratic form represents all prime numbers up to 73, then it represents all prime numbers. [Poo Sung]

The Maya, who came to prominence in the New World in the 3rd century A.D., used two calendars, based on a sacred year of 260 days and a vague year of 365 days. The least common multiple of the two calendars, called the calendar round, has 18980 days or 73 sacred years. [Dershowitz]

The success of the magazine known as *73 Amateur Radio Today* gave Wayne Green the money to launch *Byte* magazine. [Pierce]

The Discordian calendar has five 73-day seasons: Chaos, Discord, Confusion, Bureaucracy, and The Aftermath.

"My community will divide into 73 sects." Prophet Muhammad (circa 622).

The USS *George Washington* (CVN-73) currently has the largest prime hull number of any aircraft carrier in the American fleet.

79

The square root of 79 starts with four 8's. [Axoy]

The atomic prime number of gold. Precious, isn't it?

On page 79 of the novel *Contact* by Carl Sagan, it says that no astrophysical process is likely to generate prime numbers.

U.S. President James A. Garfield died 79 days after being shot. [Dowdy]

Ten to the power 79 has been called the "Universe number" because it is considered a reasonable lower limit estimate for the number of atoms in the observable universe.

There were 79 broadcasted episodes of the original *Star Trek*. [McCranie]

An Army Air Corps bomber crashed into floor 79 of the Empire State Building on July 28, 1945.

The following poetic quotation contains 79 letters: "When One made love to Zero, spheres embraced their arches and prime numbers caught their breath." Raymond Queneau (French author of the mid-20th century). [Post]

A useful estimate of $\pi(x)$ is $\mathrm{li}(x) = \int_2^x 1/(\log t)\, dt$. For small x, see Figure 25, $\mathrm{li}(x) > \pi(x)$; but in 1912 Littlewood proved this is not always the case. His student Skewes sought the least x for which $\mathrm{li}(x) < \pi(x)$ and, assuming that the Riemann hypothesis is true, showed it is less than $e^{e^{e^{79}}}$. (The current estimate for the first crossover occurs around $1.4 \cdot 10^{316}$.)

Amazon Prime^{TM}(Amazon.com's shipping club) is a 79 dollar per year service that allows you to get free two-day delivery and discounted next-day delivery on in-stock items.

The composite volcano Mt. Vesuvius erupted in A.D. 79, burying Pompeii, Herculaneum, and Stabiae under ashes and mud.

Figure 25. $\pi(x)$ (solid) verses $\mathrm{li}(x)$ (dotted)

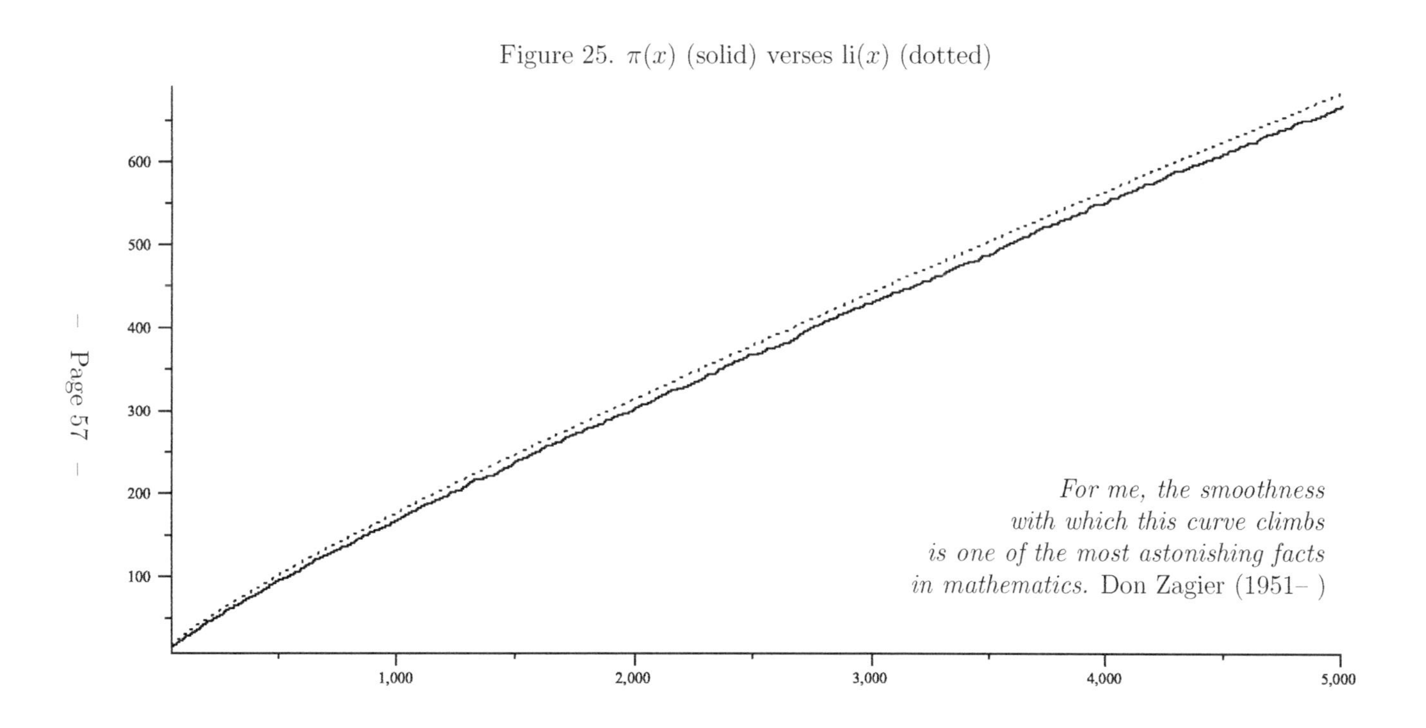

83

The smallest prime whose square, 6889, is a **strobogrammatic** number. [Wu]

In the movie *83 Hours 'Til Dawn*, an heiress is kidnapped for ransom and buried alive in a special capsule that within 83 hours will become her tomb. It is based on a true story.

Number theorist Paul Erdős lived to be 83 years old.

The number of permutations of the 10 distinct digits taken 9 at a time that are perfect squares. These range from $10124^2 = 102495376$ to $30384^2 = 923187456$. [Beiler]

The number of the French department Var is 83.

In W. H. Auden's famous poem, "Miss Gee" lived in Clevedon Terrace at number 83. [La Haye]

There are exactly 83 right-truncatable primes. [Angell and Godwin]

The first prime that can be written as a sum of composites in more ways than it can be written as a sum of primes. [Hartley]

83 is the sum of the squares of the first three consecutive odd primes ($3^2 + 5^2 + 7^2$). [Gallardo]

The smallest multidigit **curved-digit prime**. [Gupta]

Find the average of all primes up to 83 and you'll get the reversal of 83.

The exact number of Johnson Solids with no hexagonal faces. [Hartley]

The German "mental calculator" Rüdiger Gamm (1971–) once demonstrated his ability on an Australian radio show by calculating consecutive powers of 83 in his head without making a mistake. The number 83 had been selected for him randomly.

Vanuatu is an archipelago of 83 islands lying between New Caledonia and Fiji in the Southwest Pacific.

89

Two to the power 89, minus one, was first proved prime by a man whose last name was Powers.

The smallest **composite-digit prime**.

89 is the smallest **circular-digit prime**.

$2^2 + 3^3 + 5^5 + 7^7 + 11^{11} + \ldots + 89^{89}$ is prime. [Crespi de Valldaura]

$89 = 8 + 9^2$.

Hugo Steinhaus (1887–1972) proved that if you take any positive integer, find the sum of the squares of its digits, and then repeat; that you will either come to 1 or the cyclic sequence 145, 42, 20, 4, 16, 37, 58, **89**.

89 is the smallest prime (indeed the smallest positive integer) whose square (7921) and cube (704969) are likewise prime upon reversal. [Trotter]

The longest verse in the KJV Bible is Esther 8:9 with 89 plus one words.

The smallest prime for which the sum of all odd primes less than or equal to it is a square. [Astle]

The smallest **Sophie Germain prime** to start a **Cunningham chain** of length 6 (1st kind): (89, 179, 359, 719, 1439, 2879).

Hellin's law states that twins occur once in 89 births, triplets once in 89^2 births, and quadruplets once in 89^3 births, and so forth. This approximation came before the advent of fertility methods. [Crown]

Most small numbers quickly produce palindromes if you repeatedly reverse-then-add. However, 89 requires two dozen iterations, far more than any other small prime (see Table 2).

The smallest **holey prime**.

97

The last two digits of Jack Reacher's ATM card PIN in the novel *Bad Luck and Trouble* by Lee Child. Reacher liked 97 because it is the largest two-digit prime number. [Reynolds]

Table 2. An Example of the Reverse-then-Add Process (89)

89 + 98 = 187
187 + 781 = 968
968 + 869 = 1837
1837 + 7381 = 9218
9218 + 8129 = 17347
17347 + 74371 = 91718
91718 + 81719 = 173437
173437 + 734371 = 907808
907808 + 808709 = 1716517
1716517 + 7156171 = 8872688
8872688 + 8862788 = 17735476
17735476 + 67453771 = 85189247
85189247 + 74298158 = 159487405
159487405 + 504784951 = 664272356
664272356 + 653272466 = 1317544822
1317544822 + 2284457131 = 3602001953
3602001953 + 3591002063 = 7193004016
7193004016 + 6104003917 = 13297007933
13297007933 + 33970079231 = 47267087164
47267087164 + 46178076274 = 93445163438
93445163438 + 83436154439 = 176881317877
176881317877 + 778713188671 = 955594506548
955594506548 + 845605495559 = 1801200002107
1801200002107 + 7012000021081 = **8813200023188**

There are 97 chapters, some as short as a paragraph or two, in Bruce Chatwin's book *In Patagonia*, giving the entire book a vignette-like feel. [Deemikay]

The smallest odd non-cluster prime. An odd prime p is called a cluster prime if every even positive integer less than $p - 2$ can be written as the difference of two primes $q - q'$, where q, $q' \leq p$.

The number formed by the concatenation of odd numbers from one to 97 is prime. [Howell]

Shakespeare's Sonnet XCVII (97) has a curious property. The

seventh word of the seventh line of the sonnet is "prime." [Keith]

There are 97 leap days every four hundred years in the Gregorian calendar. [Poo Sung]

97, 907, 9007, 90007 and 900007 are all primes, but 9000007, 90000007, 900000007, 9000000007, and 90000000007 are all composites.

The continued fraction for π to the fourth power starts with 97.

Generally, one jigger (one and a half ounces) of liquor (gin, rum, vodka, and whiskey) contains 97 calories. [Larsen]

The first four pairs of digits following the decimal point of $\frac{1}{97}$ are powers of three: $\frac{1}{97} = 0.01030927....$ [Hill]

The first ever NASCAR win for Plymouth and the Chrysler Corporation was powered by a strictly stock flathead engine that produced 97 horsepower.

In the middle of 97 insert $97 - 1$ and you get 9967, which is also prime. Put $97 - 1$ in the middle of 9967 and you get 999667, which is another prime. Put $97 - 1$ in the middle again and you get 99996667, yet another prime. After inserting $97 - 1$ in the middle once more, it is left for the **prime curiologist** to find the factors of 9999966667. [Honaker]

$97 = 4 \cdot 4! + \frac{4}{4}$. [Dunn]

101

The smallest **smoothly undulating** palindromic prime (SUPP) and the only known SUPP of the form $(10)_n1$.

Australia's sugar industry once imported 101 cane toads from Hawaii in the hope that they would kill cane beetles threatening their sugar crop.

$101 = 5! - 4! + 3! - 2! + 1!$.

The 101 proof of the Kentucky bourbon Wild Turkey is its most common bottling.

Using the alphabet code (page 50), DIVISION is 101. [Cox]

Professor Nash's office number in the movie *A Beautiful Mind*.

A radio astronomer in the movie *Contact* makes the statement "OK, a hundred and one, the pulse sequence through every prime number between two and a hundred and one." [Haas]

Depeche Mode named one of their albums "101". [Gevisier]

The plot of Walt Disney's cartoon film *One Hundred and One Dalmatians* (often abbreviated as *101 Dalmatians*) centers on the fate of the kidnapped puppies of Pongo and Perdita.

Room 101 was the place where your worst fears were realized in George Orwell's classic *Nineteen Eighty-Four.*

Thor Heyerdahl's book *Kon-Tiki* is about his harrowing 101-day feat of crossing the Pacific on a balsa log raft which made him world famous.

Room 101 is the name of a British TV series in which celebrities are invited to discuss their (often whimsical) pet hates with the host.

In a short section titled "Prime Numbers ... and Monkeys" of Danica McKellar's bestselling book *Math Doesn't Suck*, the following statement can be found: "It's hard to believe that there's no way to evenly divide up 101, but it's true!"

The first odd prime for which the Mertens function $M(n) = 0$ (see Figure 31 on page 77). [Post]

A googol (10^{100}) contains 101 digits. The search engine *Google*'s play on the term reflects the company's mission to organize the immense amount of information available on the web.

103

The smallest prime whose reciprocal contains a period that is exactly $\frac{1}{3}$ of the maximum length. [Wells]

Proof, an award-winning play by David Auburn, revolves around a mysterious mathematical proof involving prime numbers left behind in 103 notebooks by a young woman's brilliant father.

There are exactly 103 geometrical forms of magic knight's tour of the chessboard.

M103 and NGC 103 are both open clusters of stars in Cassiopeia. [Necula]

103 divides the concatenation of the integers 103 down to 1. [De Geest]

Using a standard dartboard, 103 is the lowest possible prime that cannot be scored with two darts.

107

When Lehmer's polynomial, $n^2 - n + 67374467$, is evaluated at an integer n, the result is never divisible by any prime less than 107. [Kravitz]

Allan Brady proved in 1983 that the maximal number of steps that a four-state Turing machine can make on an initially blank tape before eventually halting is 107.

There are exactly 107 hole-free heptominoes (polyominoes made up of seven squares connected edge-to-edge; Figure 26). [Beedassy]

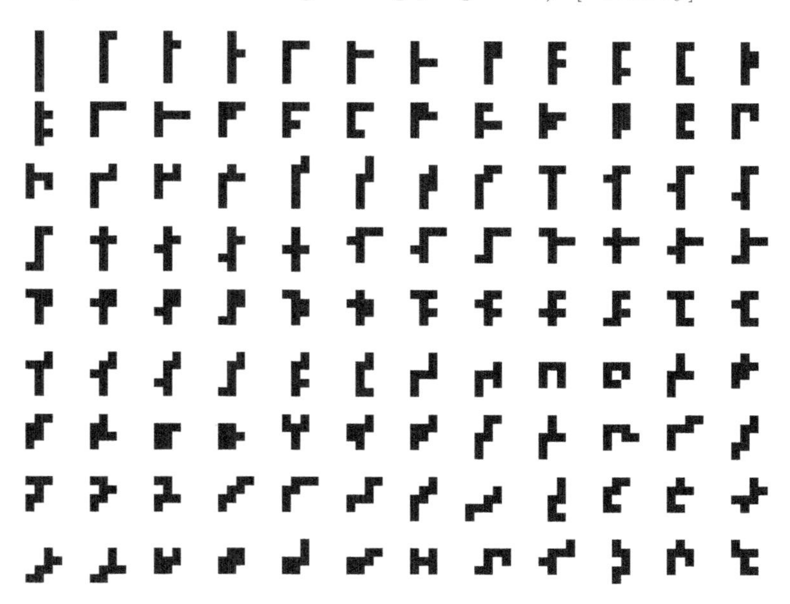

Figure 26. The Heptominoes (107 Without Holes)

The first three-digit Mersenne prime exponent.

2^{107} reversed and $2^{107} - 1$ are primes. [Nash]

Rudy Giuliani was the 107th mayor of New York City. [Trotter]

Space Shuttle Columbia (STS-107 mission) disintegrated during reentry. [Gupta]

The sum of atomic numbers of five chemical elements that have their symbols embedded in the second author's last name (H,O,Na,K,Er). [Necula]

The Peugeot 107 is a city car produced by the French automaker Peugeot. (It has proved to be quite popular with British buyers.)

"One hundred and seven" is the smallest positive integer requiring six syllables (in English) if 'and' is included.

109

If 109 is written in Roman notation (CIX), then it becomes reflectable along the line it is written on.

The pipe organ at the Cathedral of Notre Dame in Paris has 109 stops.

The Caldwell Catalog contains 109 deep-sky delights for backyard stargazers.

$103\# + 107$ is divisible by 109. Note the use of consecutive primes. [Rupinski]

109 equals the square root of 11881 or $118 - 8 - 1$.

The registry number of the famous patrol boat PT-109, commanded by John F. Kennedy.

There are 109 letters in the following quotation: "Given the millennia that people have contemplated prime numbers, our continuing ignorance concerning the primes is stultifying." R. Crandall and C. Pomerance, from *Prime Numbers: A Computational Perspective*, Springer-Verlag, 2001. [Post]

The Sun is just over 109 times the diameter of the Earth. [Friedman]

Shibuya 109 is a trend setting fashion complex for young women in Tokyo, Japan. One of the buildings there is called "The Prime." [Anansi]

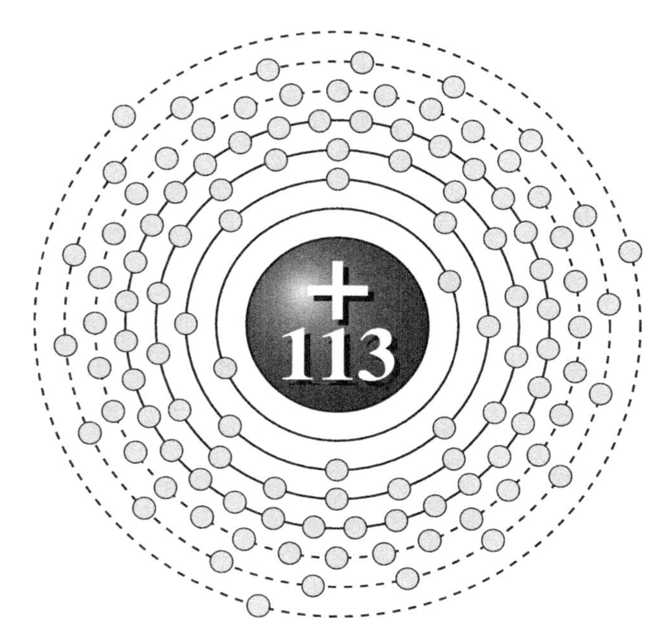

Figure 27. Electronic Configuration of Ununtrium (Uut)

113

The smallest three-digit **absolute prime**. [Richert]

Zu Chongzhi (or Tsu Ch'ung Chi), along with his son Zu Gengzhi, stated in a mathematical text titled *Zhui Shu* (Method of Interpolation) that π is approximately three hundred fifty-five divided by 113.

$113^2 = 12769$ and its reversal $96721 = 311^2$. [Friend]

The atomic number of an element temporarily called ununtrium (symbol Uut, Figure 27). First created in labs in 2003, only a few atoms of this radioactive metal have ever been produced. [Kulsha]

113 is the sum of two consecutive squares: $7^2 + 8^2$. [Schlesinger]

(Joke) Theorem: Any number is prime. Proof: by the alphabet code (page 50), the phrase "any number" is 113. [Gallardo]

The Police (and general emergency) telephone number in Italy.

The only solution (in positive integers) to $a^2 + b^3 = c^7$ is $15312283^2 + 9262^3 = \mathbf{113}^7$.

Longitude 113 degrees West passes through Dolphin Island in Great Salt Lake, Utah. Think of it as a "prime meridian" for the world's saltiest sailors.

The number of teeth on each of the Photoelectric Sieve gears was a multiple of one of the primes up to 113 (see Figure 28).

Let $\psi(n)$ be the natural logarithm of the least common multiple of all the numbers less than or equal to n (so $\psi(6) = \log(60)$). The maximum value of $\frac{\psi(n)}{n}$ occurs at 113.

127

The smallest odd prime number that cannot be expressed as the sum of a power of two and a prime (so it is also the smallest prime that becomes composite upon changing any one of the bits in its binary expansion). [Capelle]

In the 19th century, Camille Armand Jules Marie, better known as the "Prince de Polignac," missed the fact that 127 is not an odd number that is the sum of a power of two and a prime. Andy Edwards coined the name "obstinate numbers" to describe such integers.

There are 127 prime pairs that sum to ten thousand. [Richstein]

$127 = -1 + 2^7$. (All digits are used in the same order.) [Jeursen]

$2^0 + 2^1 + 2^2 + 2^3 + 2^4 + 2^5 + 2^6 = 127$.

127 can also be expressed as the sum of factorials of the first three odd numbers $(1! + 3! + 5!)$.

John Barrymore kissed Mary Astor and Estelle Taylor a total of 127 times in the film *Don Juan* (1926). [Dobb]

The last three digits of the eleventh Mersenne prime is 127. Note that the next Mersenne prime happens to be $M(127)$. [Honaker]

127 millimeters is exactly five inches. [Vrba]

Figure 28. Photoelectric Sieve

This Photoelectric Sieve was built by Dr. Derrick Henry Lehmer in 1932 and became operational the following year at the World's Fair in Chicago. Holes adjacent to each gear tooth that did not correspond to modular solutions of the problem being solved were plugged with toothpicks. A bright light was then set on one side of the gears, and a photoelectric cell was placed at the other end to detect when the holes all lined up. Using a vacuum-tube amplifier to multiply the strength of the resulting signal by 700,000,000, the device could test 300,000 numbers per minute and was used to complete the factorization of $2^{79} - 1$ in a matter of seconds. Because of the high amplification, it was sometimes inadvertently "jammed" by local ham radio transmissions.

ASCII characters (the long-established way to represent characters on a computer) are numbered from 0 to 127.

The number of prime-numbered days of the month in a leap year. [Cabisco]

52	61	4	13	20	29	36	45
14	3	62	51	46	35	30	19
53	60	5	12	21	28	37	44
11	6	59	54	43	38	27	22
55	58	7	10	23	26	39	42
9	8	57	56	41	40	25	24
50	63	2	15	18	31	34	47
16	1	64	49	48	33	32	17

Figure 29. A Franklin Square (127 Equations!)

Maya Mohsin Ahmed of UC Davis found that the numbers in an 8-by-8 *Franklin square* (Figure 29) can be described by 127 equations. When Benjamin Franklin (1706–1790) wasn't flying kites, the noted polymath found time to experiment with recreational mathematics. Among other things, he invented magic square variants that are constructed with nonnegative numbers and contain the following properties: the entries of each row and column add to a common (or magic) sum; half of each row or column sums to half of the magic sum; the four corner entries together with the four middle entries add to the magic sum; in addition, each of the "bent rows" (as Franklin called them) have the magic sum. We still do not know what method Franklin used to construct his squares, and leave it for the reader to find other interesting properties.

131

The 32nd prime. Note that $1 + 3 + 1 = 3 + 2$.

The nematode *Caenorhabditis elegans* hermaphrodite has exactly 131 cells that are eliminated by programmed cell death (apoptosis). [Haga]

A rare prime of the form $2^p + 3$, where p is prime. [Luhn]

Locomotive number 131 plunges off the edge of a ravine in the sci-fi movie *Back to the Future Part III.*

The sum of $2 + 3 + 4 + \ldots + 19$ minus the sum of primes less than 19. [Trotter]

Iodine-131 is a radioactive isotope used in thyroid disease diagnosis and therapy. [Barnhart]

The sum of the first 131 non-primes is prime. [Patterson]

A prime Ulam number that is the sum of two consecutive Ulam numbers ($62 + 69 = 131$). Can you find a greater prime example?

The smallest mountain prime, i.e., a prime that satisfies the following conditions: the end digits are 1; the first digits are in strictly increasing order and the last digits are in strictly decreasing order; there is only one largest digit. [Pol]

The largest prime factor of the number of mountain primes ($2620 = 2^2 \cdot 5 \cdot 131$).

137

The start of twelve consecutive primes with symmetrical gaps about the center.

The sum of Rosie's measurements (42–39–56) equals 137 in the song "Whole Lotta Rosie" by the Australian hard rock band AC/DC.

137 is the largest prime factor of 123456787654321.

William Shanks (1812–1882), a British amateur mathematician, manually calculated the logarithms of 2, 3, 5, and 10 to 137 decimal places in 1853.

The Hawaiian Island chain is made up of 137 islands, islets, and shoals.

A molecule of chlorophyll *a*, $C_{55}H_{72}MgN_4O_5$, consists of 137 atoms (see Figure 30). [Blanton]

The numerical value of the Hebrew word *Kabbalah* is 137 (using the common Hebrew gematria method), which is equal to the combined

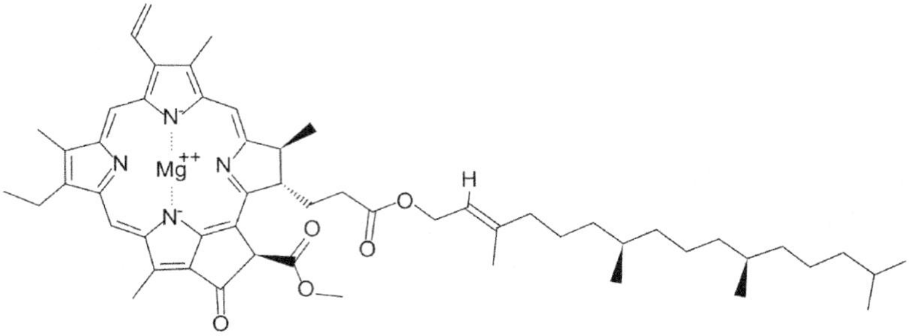

Figure 30. The Structure of Chlorophyll *a*'s 137 Atoms

value of the words chochmah (73), "wisdom," and nevuah (64), "prophecy." [Scholem]

One of the math papers of Ted Kaczynski (also known as the Unabomber) is *Boundary Functions for Bounded Harmonic Functions*, Transactions of the American Mathematical Society 137.

Unabomber Sketch

The reciprocal of the fine-structure constant of electromagnetism is close to 137. Throughout his life, physicist Wolfgang Pauli had been preoccupied with the question of why the fine-structure constant has this value. He died in Room 137 of the Rotkreuz hospital in Zürich, Switzerland.

"And these are the years of the life of Ishmael, an hundred and thirty and seven years: and he gave up the ghost and died; and was gathered unto his people." (Genesis 25:17, KJV)

The sum of the squares of the digits of 137 is another prime and all five odd digits are used. [Trotter]

The James A. FitzPatrick nuclear power plant on the shore of Lake Ontario has 137 control rods. [Gonyeau]

137 is the Chilean National Sea Rescue emergency number. [Vogel]

Mabkhout (1993) showed that every number $x^4 + 1$, for $x > 3$, has a prime factor greater than or equal to 137. He used a classical result of Størmer. [Post]

The only known **primeval number** whose sum of digits equals the number of primes "contained."

139

139 divides the sum of the first 139 composite numbers. [Honaker]

139 and 149 are the first consecutive primes differing by 10. [Wells]

$3^{139} + 2$ is prime.

The smallest prime that contains one prime digit, one composite digit, and one digit that is neither prime nor composite. [Brown]

The smallest prime factor of the smallest multidigit composite **Lucas number** with a prime index.

The number of chromatically unique simple graphs on seven nodes. [Post]

STB 139 is the cryptic name for a nightspot located in the Rippongi district of Tokyo, Japan.

In 1975, Pomerance showed that the second largest prime factor of an odd **perfect number** should be at least 139.

The first odd prime that appears in Stirling's series for $n!$ is 139.

$$n! = \sqrt{2\pi n}\left(\frac{n}{e}\right)^n \left(1 + \frac{1}{12n} + \frac{1}{288n^2} - \frac{\mathbf{139}}{51840n^3} - \frac{571}{2488320n^4} + \cdots\right)$$

149

An emirp formed from the digits of first three perfect squares. [Gupta]

Elvis Presley had no less than 149 songs to appear on the Billboard Hot 100 popularity chart in the United States. [Blanchette]

There are 149 ways to put 8 queens on a 7-by-7 chessboard so that each queen attacks exactly one other queen. [Gardner]

$149 = 6^2 + 7^2 + 8^2$. [Schlesinger]

The only known prime in the concatenate square sequence. There are no others within the first 14916 terms. Note that a prime in the concatenate cube sequence has yet to be found.

The smallest emirp with no prime digits. [Beedassy]

151

The smallest palindromic prime occurring between two consecutive squareful (or non-squarefree) numbers. [Gupta]

U.S. Route 151 is an important diagonal highway that runs northeasterly through the states of Iowa and Wisconsin. [Enslin]

Preston Trucking called itself "The 151 Line." [Litman]

Bacardi 151 Rum is a high proof, dark, aromatic rum. [Street]

The total number of types of Pokémon in the original set (Bulbasaur to Mew, not counting the glitch Pokémon Missingno).

The smallest palindromic prime that is not a Chen prime. [Melik]

Psalm 151 is not found in the traditional Hebrew text of the Jewish Bible.

157

The start of the smallest string of consecutive **equidigital numbers** of length seven. [Santos and Pinch]

$\frac{157^{157}+1}{157+1}$ is prime. [Wagstaff]

The largest odd integer that cannot be expressed as the sum of four distinct nonzero squares with greatest common divisor 1.

The smallest prime of the form $2^p + p^3$, where p is prime. [Brown]

Two to the power 157 is the smallest "apocalyptic number," i.e., a number of the form 2^n that contains '666'. [Pickover]

Samuel Yates, who did an extensive work on the topic of prime period lengths, lived at 157 Capri-D Kings Point in Delray Beach, Florida.

157 is the smallest three-digit prime that produces five other primes by changing only its first digit: 257, 457, 557, 757, and 857. [Opao]

Bus number 157 is hijacked in the Clint Eastwood thriller *Dirty Harry*. [May]

The most commonly identified Shiga toxin-producing *Escherichia coli* in North America is *E. coli* O157.

157^2 and $(157+1)^2$ use the same digits.

163

A prime whose reversal is another prime (19) squared. [Trigg]

The largest Heegner number, i.e., the largest integer d such that the imaginary quadratic field $\mathbb{Q}(\sqrt{-d})$ has unique factorization (class number 1). The others are 1, 2, 3, 7, 11, 19, 43, and 67. [Croll]

The phrase "is a prime number" sums to 163 using the alphabet code (page 50). [Necula]

In the April 1975 issue of *Scientific American*, Martin Gardner wrote (jokingly) that Ramanujan's constant , $e^{\pi\sqrt{163}}$, is an integer. It is very close!

$$e^{\pi\sqrt{163}} = 262537412640768743.99999999999925...$$

The name "Ramanujan's constant" was actually coined by Simon Plouffe and derives from the above April Fool's joke played by Gardner. The French mathematician Charles Hermite (1822–1901) observed this property of 163 long before Ramanujan's work on these so-called "almost integers." [Aitken]

167

Wieferich proved that 167 is the only prime requiring exactly eight cubes to express it. [Rupinski]

Indonesia has 167 volcanoes.

In 2003, outgoing Governor George Ryan of Illinois commuted the death sentences of 167 death row inmates two days before leaving office, calling the death penalty process "arbitrary and capricious, and therefore immoral."

At the start of backgammon each player has a pip count of 167. [La Haye]

The number of prime quadruples, counting $(3, 5, 7, 11)$, below one million. (A prime quadruple is four consecutive primes, such that the first and the last differ by 8. It has the form $(p, p+2, p+6, p+8)$ for $p > 3$).

The least prime factor of the smallest vampire number that has prime fangs of the same length ($117067 = 167 \cdot 701$). A vampire number is an integer which can be written as the product of two factors (its "fangs") whose digits together are a rearrangement of the original number.

The smallest number whose fourth power begins with four identical digits.

A highly cototient number is an integer $k > 1$ with more solutions to $x - \phi(x) = k$ than any other integer below k and above one. Here $\phi(x)$ is Euler's totient function: the number of positive integers less than or equal to x that are relatively prime to x. All highly cototient numbers greater than 167 are congruent to 9 (mod 10).

173

There are 173 stars included in the list near the back of the *Nautical Almanac*, an annual publication of the U.S. Naval Observatory.

The destroyer escort USS *Eldridge* (DE-173) is made invisible and teleported from Philadelphia, Pennsylvania, to Norfolk, Virginia, in an alleged 1943 incident known as the Philadelphia Experiment. [Moore]

The smallest prime inconsummate number, i.e., no number is 173 times the sum of its digits. [Conway]

179

A winning solution to the 15-hole triangular peg solitaire game is: (4,1), (6,4), (15,6), (3,10), (13,6), (11,13), (14,12), (12,5), (10,3), (7,2), (1,4), (4,6), (6,1). The term (x,y) means move the peg in hole x to y. Not only does this solution leave the final peg in the original empty hole, but the sum of the peg holes in the solution is prime.

$179 = (17 \cdot 9) + (17 + 9)$. [Capelle]

There are 179 even days in one year. [Kik]

Hafnium-179 is the stable nuclide with the largest quadrupole moment.

Bobby Fischer issued a multipage document with 179 demands to the World Chess Federation (FIDE) before defaulting his title to Anatoly Karpov in 1975. [Edmonds and Eidinow]

181

The sum of the first 181 primes minus 181 is prime. [Gevisier]

The 181st prime is congruent to 1 (mod 181). [Haga]

181 pennies currently minted in the U.S. weigh within a penny of a pound.

The smaller prime in the first multidigit "Gridgeman pair." A Gridgeman pair is two palindromic primes, which differ only in that their middle digits are x and $x + 1$ respectively. These are named in honor of Norman T. Gridgeman (1912–1995) who conjectured that there are an infinite number of primes in this form. [King]

191

A prime number of spaces are formed if we choose $n = 191$ distinct points on a circle so that no three of the chords that join them are concurrent. In general, the number is

$$g(n) = (n^4 - 6n^3 + 23n^2 - 18n + 24)/24.$$

(See Moser's circle problem on page 36.)

The reversal of fib(191) is prime.

The smallest palindromic prime p such that neither $6p - 1$ nor $6p + 1$ is prime. [Necula]

American Airlines Flight 191, which crashed on May 25, 1979, remains the deadliest airplane accident on U.S. soil in terms of the total number of aircraft occupants killed.

The values of the most popular United States coins currently in circulation (silver dollar, half-dollar, quarter, dime, nickel, and penny) sum to 191 cents. [Patterson]

The smallest multidigit palindromic prime that yields a palindrome when multiplied by the next prime: $191 \cdot 193 = 36863$. [Russo]

A palindromic prime whose square is a distinct-digit number whose first two digits, central digit, and last two digits, are perfect squares.

There are 191 orientable octahedral manifolds. [Beedassy]

193

The only odd prime p known for which 2 is not a primitive root of $4p^2 + 1$. (A primitive root of a prime p is an integer which has multiplicative order $p - 1$ modulo p.)

193 can be written as the difference between the product and the sum of the first four primes. [Poo Sung]

197

The sum of the first dozen primes. [Brod]

The only three-digit prime repfigit number. [Trotter]

197 is the sum of digits of all two-digit primes. [Gallardo]

The first three digits of the prime $197 \cdot 197! + 1$ are 197. [Luhn]

199

The smallest number with an additive persistence of 3. The persistence of a number is the number of times one must apply a given operation (adding the digits in this case) to an integer before reaching a fixed point: $199 \rightarrow 19 \rightarrow 10 \rightarrow 1 \rightarrow 1 \ldots.$ [Gupta]

The Cape May Lighthouse in New Jersey has 199 steps in the tower's cast iron spiral staircase. [McCranie]

The smallest prime that is the sum of the squares of four distinct primes. [Sladcik]

The smallest three-digit prime Lucas number.

199 sets a new record low for the Mertens function $M(n) = \sum_{k=1}^{n} \mu(k)$, where $\mu(k)$ is the Möbius function (see Figure 31).

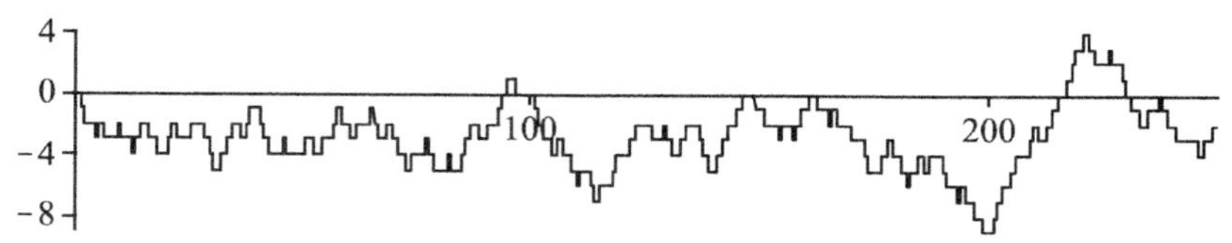

Figure 31. Graph of the Mertens Function

211

The smallest prime formed from the reverse concatenation of three consecutive Fibonacci numbers. [Gupta]

The number of primes that can appear on a 24-hour digital clock (00:00 up to 23:59). [De Geest]

G. H. Hardy once sent a postcard to his friend Ramanujan with a list of six New Year's resolutions beginning: (1) prove the Riemann hypothesis; (2) Make 211 not out in the fourth innings of the last Test Match at Oval; (3) ....

211 is a prime lucky number and there are 211 prime lucky numbers less than 10^{2+1+1}. [Post]

There are 211 stairs going from the Founder's Room to the bell chamber at Singing Tower, near Lake Wales, Florida.

Virtually all e-mail is sent using the Simple Mail Transfer Protocol (SMTP). The smallest SMTP reply code is 211 which means `system status`, or `system help reply`.

The smallest prime between 2 sets of 11 consecutive composites. Note the concatenation of 2 and 11. [Opao]

2-1-1 is an easy-to-remember telephone number that connects people with important community services and volunteer opportunities in the United States: `http://www.211.org/`. [Haga]

223

The number of primes and the number of composites that cannot be written as the sum of two primes, up to 223, are equal. [Honaker]

A chicken and human have 223 enzymes of identical sequence length. [Jolly]

The prime factors of $2^p - 1$ are all of the form $2kp + 1$, where k is a positive integer, and p is an odd prime. Fermat used this fact to discover that 223 divides the Mersenne number $M(37) = 137438953471$.

227

The smallest three-digit prime that is changed into a composite number if any digit is deleted.

"227" is an American sitcom starring Marla Gibbs that originally aired from September 1985 to May 1990. [Hultquist]

22/7 has been used for over 2000 years to approximate π. It is the first convergent $3 + 1/7$ of π's infinite continued fraction.

$$\pi = 3 + \cfrac{1}{7 + \cfrac{1}{15 + \cfrac{1}{1 + \cfrac{1}{292 + \cfrac{1}{1 + \ddots}}}}}$$

The next convergent, 355/113, is a far better approximation!

In the novel *Life of Pi* by Yann Martel, the protagonist (Pi) survives 227 days shipwrecked at sea with a Bengal tiger.

229

The smallest prime that remains prime when added to its reversal. [Luhn]

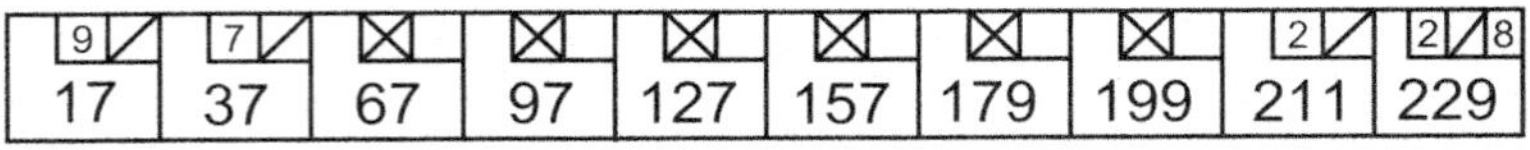

9 /	7 /	X	X	X	X	X	X	2 /	2 / 8
17	37	67	97	127	157	179	199	211	229

Figure 32. Bowling a Prime in Every Frame

The highest possible score in a standard game of bowling if your score in each of the ten frames is required to be a prime number (Figure 32). (See also the entry for 293.) [Keith]

The sum of the first 229 primes divides the product of the first 229 primes.

229 is the difference between 3^3 and 4^4. [Raymond]

Replacing each digit of prime 229 with its square, respectively its cube, results in two new primes (4481 and 88729) with a palindromic difference of 84248. Coincidentally, 229 + 4481 + 88729 is palindromic as well. [De Geest]

233

Describing 233 and repeating the process with each new term produces five more primes, i.e., "one 2, two 3's," generates 1223, etc. [Rivera]

Neil J. A. Sloane, editor-in-chief of the *On-Line Encyclopedia of Integer Sequences*, works at AT&T Shannon Laboratories in room C233.

The ratio between the adjacent Fibonacci numbers 233 and 144 is approximately the golden ratio: $\phi = \frac{1+\sqrt{5}}{2} = 1.61803398874989\ldots$. This is because the nth Fibonacci number is $\frac{\phi^n}{\sqrt{5}}$ rounded to the nearest integer.

The only known multidigit Fibonacci prime whose digits are all Fibonacci primes. [Gupta]

"Pascal's Wager," an argument from game theory that we should believe in God, appeared posthumously in Pensées 233.

Pascal (1623–1662)

Had Ray Bradbury used the metric system, he may have called his novel "Celsius 233." [Hartley]

Claimed to be the first book with no verbs, *The Train from Nowhere* by Michel Thaler (a pseudonym) has 233 pages. [Opao]

233 is the final chapter of *The Curious Incident of the Dog in the Night-Time* by Mark Haddon, which uses all prime numbers for its chapters.

239

The smallest prime factor of the palindrome 1234567654321. Note that 239 is the smallest prime with period length 7 (the palindrome's middle digit). [Yates]

In 1706, Machin discovered that $\frac{\pi}{4} = 4 \arctan \frac{1}{5} - \arctan \frac{1}{\mathbf{239}}$. Shanks used this to calculate π to 707 places in 1873, but only the first 527 digits were correct.

A solution to $x^4 + 104^4 = 58136^2 + 1$. [McLean]

Magic Square Lexicon: Illustrated by H. D. Heinz and J. R. Hendricks, includes 239 terms associated with magic squares, cubes, tesseracts, stars, etc.

A total cholesterol level of over 239 milligrams per deciliter is considered to be high.

The largest number that cannot be written as the sum of at most eight positive cubes (page 30):

$$239 = 5^3 + 3^3 + 3^3 + 3^3 + 2^3 + 2^3 + 2^3 + 2^3 + 1^3.$$

239 requires nineteen fourth powers to represent it. There is no known number which requires more! [Deshouillers]

In a breeder reactor, uranium nuclei absorb neutrons to produce plutonium-239.

Professor Dick Solomon, lead character on the TV show *Third Rock from the Sun*, teaches his classes in room number 239. [Rupinski]

One saros cycle (the time for Earth, Moon, and Sun to return to the same relative geometry) is almost exactly 239 anomalistic months (the perigee to perigee time for the Moon). [McCranie]

In *The Simpsons* episode "Homer's Night Out," Homer weighs himself at 239 pounds and resolves to exercise.

HAKMEM, alternatively known as AI Memo 239, is a technical report of the MIT AI Lab that describes a wide variety of hacks, primarily useful and clever algorithms for mathematical computation.

The book *Pandolfini's Endgame Course: Basic Endgame Concepts Explained by America's Leading Chess Teacher* consists of 239 specific endgame positions, progressing from elementary endings to some subtle minor piece and pawn situations.

K. 239 (Serenata Notturna) is Mozart's only work for two orchestras. [Schroeppel]

Saint Petersburg Lyceum 239 is a public high school in Russia that specializes in mathematics and physics. Among its famous alumni is Yuri Matiyasevich, who solved Hilbert's tenth problem.

241

The smallest prime p such that p^7 can be written as the sum of 7 consecutive primes. Note that $2 + 4 + 1 = 7$. [Rivera]

Internationally renowned researcher Dr. Karen Rogers did a study of 241 profoundly gifted children in 1994–1995 during a postdoctoral fellowship. It was especially useful to parents who had observed developmental differences in their children but were unaware of what those differences may signify.

The radioisotope americium-241 is used in many smoke detectors.

The smallest prime p such that p plus the reversal of p equals a palindromic prime. [Luhn]

The prime numbers are identified as N0241 in Sloane's original "Handbook of Integer Sequences."

251

The smallest number that is the sum of three cubes in two ways: $1^3 + 5^3 + 5^3 = 2^3 + 3^3 + 6^3$.

The sum of the letters of "two hundred and fifty one" is the only self-described prime if we use the alphabet code (page 50). [Lundeen]

251 can be expressed using the first three primes: $2^3 + 3^5$. [Sladcik]

The smallest of four consecutive primes in arithmetic progression (251, 257, 263, and 269). [Gallardo]

Members of the Vermont 251 Club attempt to visit 251 cities and towns in the state of Vermont.

257

The largest known prime of the form $n^n + 1$. It is very likely that 2 and 5 are the only others.

Lab 257, a book by Michael C. Carroll, is (the subtitle asserts) "the disturbing story of the government's secret Plum Island germ laboratory."

The smallest three-digit prime with distinct prime digits. [Moore]

The smallest odd octavan prime, i.e., of the form $p = x^8 + y^8$. [Russo]

The smallest prime of the form $128k + 1$. Note that any prime factor of F_n, where $n > 2$, is of this form.

The largest prime in a sequence of fifteen primes of the form $2t + 17$, where t runs through the first fifteen triangular numbers, i.e., positive integers of the form $\frac{n(n+1)}{2}$. [Silva]

263

The largest known prime whose square is strobogrammatic.

Anne Frank (and her family) hid from the Nazis in an annex behind the house at Prinsengracht 263 in Amsterdam.

The number of digits in $263!_2$ (double factorial) is 263. No other prime shares this property. [Firoozbakht]

The smallest prime formed by inserting a semiprime between the semiprime's factors.

The middle prime in the only set of three 3-digit primes that contain all of the digits 1 through 9, and whose sum is a 3-digit number. [Gardner]

In August 2007, mathematician Ali Nesin was jailed under Article TCK 263 of Turkish Criminal Code (running an illegal educational institution) for conducting a mathematics summer camp.

269

The Electoral College vote for President of the United States could end in a 269–269 vote tie. The race would then go to the House of Representatives where each state delegation would get one vote. [Patterson]

The longest official game of chess on record (269 moves) took place in Yugoslavia on 2/17/89 and ended in a draw. Note that 2, 17, 89, and 269 are all prime numbers.

271

A prime formed from the first three digits of the base for natural logarithms: $e = 2.7182818284\ 5904523536\ 0287471352....$

George W. Bush won the U.S. Presidency with 271 electoral votes in the 2000 Presidential Election, yet Al Gore had received a majority of the popular vote.

HMS *Plym* (K 271), a River-class anti-submarine frigate built for the Royal Navy was vaporized in Britain's first A-bomb explosion off the Monte Bello Islands, Australia, on the day tea rationing was lifted in the United Kingdom (October 3, 1952). [Croll]

2^{271} reversed is prime. [Nash]

271^2 and 271^3 form primes upon reversal. [Trotter]

There are exactly 271 positive numbers that give larger numbers when you write out their English names and add the letters using the alphabet code (page 50). [Hartley]

The smallest prime p such that $p - 1$ and $p + 1$ are each divisible by a cube greater than one. [Beedassy]

The number of possible bowling games with a score of 271 is the next prime. (Two bowling games are the same if the number of pins knocked down each roll are equal.)

277

The sum of the letters of "Holy Bible" if we use the alphaprime code: a = 2, b = 3, c = 5, d = 7, e = 11, ... (The Pentateuch was translated into Greek, circa 277 B.C.) [Zirkle]

In 1996, Dolly became the first animal ever to be successfully cloned from an adult somatic cell by nuclear transfer and was the only survivor of 277 cloning attempts! (The sheep was cloned from a mammary gland cell and named after the busty country music artist Dolly Parton.)

The sum $\frac{1}{2} + \frac{1}{3} + \frac{1}{5} + \frac{1}{7} + \frac{1}{11} + \ldots + \frac{1}{271} + \frac{1}{277}$ just exceeds the first prime number. [Wilson]

Emily Dickinson composed 277 poems in 1862. [Szegedy-Maszak]

The Grand Canyon is 277 miles long. [Aldridge]

The smaller prime factor of semiprime 9000007, the first composite number of the series 97, 907, 9007, etc.

A primary pretender for the base b is the smallest composite number n for which $b^n \equiv b \pmod{n}$. The list of these (when $b = 0, 1, 2, \ldots$) begins

4, 4, 341, 6, 4, 4, 6, 6, 4, 4, 6, 10, 4, 4, 14, 6, 4, 4, 6, 6, 4,
4, 6, 22, 4, 4, 9, 6, 4, 4, 6, 6, 4, 4, 6, 9, 4, 4, 38, 6, 4, 4, ...

Do you see the pattern yet? There are only 132 distinct numbers that appear in this list, but it repeats with a period of length $23\# \cdot 277\#$. (Take a moment and calculate how large that period is!)

281

Wilfred Whiteside of Houston, Texas, discovered a 7-by-7 array in which 281 primes can be found (Figure 33). It was discovered on April 29, 1999.

3	1	3	7	3	3	9
9	9	2	3	3	3	3
6	9	7	7	8	9	4
7	6	1	5	9	1	9
7	7	3	4	2	1	1
9	9	4	7	9	3	9
3	3	7	1	9	9	9

Figure 33. 281 Primes

283

The smallest prime factor of the first composite numerator of a Bernoulli number. The Bernoulli numbers B_k can be recursively defined by setting $B_0 = 1$, and then using, for $k > 0$,

$$\binom{k+1}{0}B_0 + \binom{k+1}{1}B_1 + \cdots + \binom{k+1}{k}B_k = 0.$$

They are closely related to the values of the **Riemann zeta function** at negative integers.

$283 = (6! - 5! - 4! - 3! - 2! - 1! - 0!)/2$. [Vatshelle]

293

The number of ways to make change for a dollar using the penny, nickel, dime, quarter, half-dollar, and dollar.

The largest possible prime bowling score. [Patterson]

202^{293} begins with the digits 293 and 293^{202} begins with the digits 202. [Hartley]

In the Christmas classic *It's A Wonderful Life*, George's guardian angel (Clarence Oddbody) says he will be 293 next May.

The sum of the first three **tetradic primes**: $11 + 101 + 181 = 293$. [Post]

307

The square of 307 is palindromic.

In the movie *The Odd Couple*, when Felix checks into a hotel, the first room the desk clerk tries to assign him is 307. [Litman]

311

A right-truncatable prime that can be obtained by concatenating the first three non-composite digits of π. [Necula]

Band 311 (pronounced "three eleven") got their name from a police code after one of the band's former members was charged with indecent exposure. [Puckett]

311 is the smallest number expressible as the sum of consecutive primes in four ways. [De Geest]

Apartment number 311 is the first door to be knocked on in the movie *Ace Ventura: Pet Detective*. [Shadyac]

313

If 313 people are chosen at random, then the probability that at least five of them will share the same birthday is greater than 50%.

The paper *Light and Number: Ordering Principles in the World of an Autistic Child* by Park and Youderian begins on page 313 in the *Journal of Autism and Childhood Schizophrenia* (1974). Note that the Park's child Ella was fascinated by the "order" of numbers, especially primes.

There are 313 exclamation marks in the KJV Bible! The first one occurs at the end of Genesis 17:18. [Bennet]

The smallest positive integer solution to $313(x^3 + y^3) = z^3$ contains numbers that are titanic.

Ever since Donald Duck's car first appeared in 1938 (a "1934 Belchfire Runabout" that he built from spare parts), it has frequently sported the license plate number 313. [Haga]

The largest known prime that divides a unitary perfect number, i.e., an integer which is the sum of its proper unitary divisors, not including the number itself (see *divisor* on page 254).

The telephone area code serving the city of Detroit, Michigan. It was popularized in the hip hop drama film *8 Mile* starring Eminem. [Anon]

Islam's first battle (The Battle of Badr) against the pagans of Mecca was fought and won by 313 Muslims. In Shi'ism, it is believed that Imam al-Mahdi (the ultimate savior of the believers) will appear when there are 313 true and sincere Shia followers in the world. [Daniyal]

As part of the challenges Fermat started (page 50), Frenicle challenged Wallis to solve $x^2 - 313y^2 = 1$. Wallis (and others) quickly found the smallest nonzero solution was $x = 32188120829134849$ and $y = 1819380158564160$. [Beiler]

$313 = 12^2 + 13^2$. [De Geest]

In the movie *Somewhere In Time*, playwright Richard Collier (Christopher Reeve) checks into room number 313 of the Grand Hotel on Mackinac Island, Michigan. [Haga]

The Roman emperor Constantine made Christianity the official religion in A.D. 313. [Pitts]

313 (base ten) = 100111001 (base two). Note that 100111001 (base ten) is a palindromic prime as well. The only three-digit number with this property. [Larsen]

Eris, the largest known dwarf planet in the solar system, was previously designated UB_{313} in 2003. (It is larger than Pluto.)

The smallest number to appear exactly three times in its own factorial (313!). [ten Voorde]

313 is the name of a unique wakeboarding trick: "Heelside Raley with a frontside handle-pass 360." [Hill]

$313 \mapsto 3 \cdot 13 = \underline{3} + 5 + 7 + 11 + \underline{13}$. [Vrba]

317

The number of ones that form the fourth repunit prime. It was first proven prime by Hugh C. Williams in 1977. [Dobb]

"317 is a prime, not because we think so, or because our minds are shaped in one way rather than another, but *because it is so*, because mathematical reality is built that way." (G. H. Hardy, *A Mathematician's Apology*, 1940)

Romanian mathematician Dimitrie Pompeiu (1873–1954) posed the following puzzle: ABC398246 is a nine-digit number exactly divisible by 317, whose first three digits (A,B,C) are unknown. What are the digits A, B, and C?

$317 = (-3)^3 + 1^3 + 7^3$. [Dobb]

A door with room number 317 appears in film footage taken of Lee Harvey Oswald shortly before his death.

331

Robert Franek wrote a book entitled *The Best 331 Colleges.*

The total population of Rives, Tennessee (the address of one of the authors), as of the 2000 census.

133 331
1333 3331
13333 33331
133333 333331
1333333 3333331
13333333 33333331
composite and prime

Sound travels just over 331 meters per second in dry (0% humidity) air at 0 °C. The approximate speed (at one atmosphere) can be calculated from the equation

$$c_{\text{air}} = 331.3\sqrt{1 + \frac{\vartheta}{273.15}} \quad \text{meters per second,}$$

where ϑ is the temperature in Celsius. [Necula]

337

The smallest prime formed from the concatenated factors of a repunit $(3 \cdot 37)$.

$81^3 + 162^3 = 9^7$. Note the exponents. [Rubin]

$337 = 2^{2^3} + 3^{2^2}$. [Kulsha]

The sum of the areas of the rectangles in this Fibonacci paradox: if we divide a 13-by-13 square into four pieces as shown in Figure 34, it appears that they can be rearranged into the one unit smaller 8-by-21 rectangle.

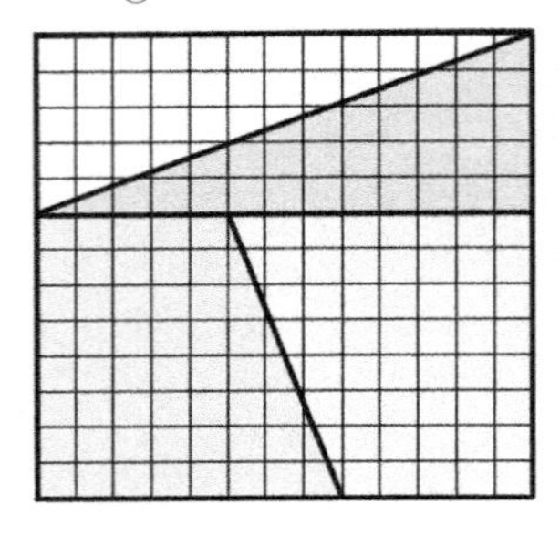
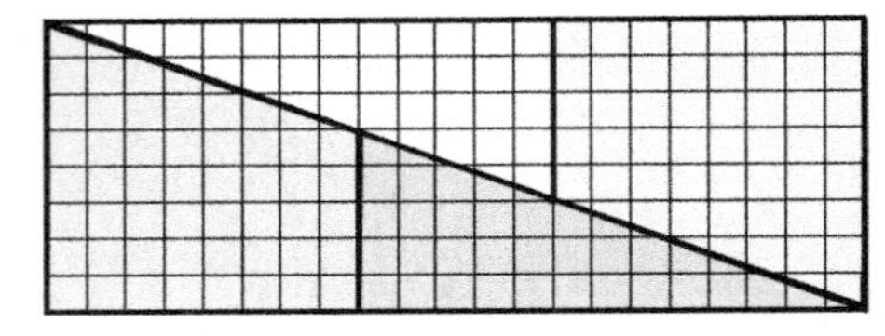

Figure 34. A Fibonacci Dissection Paradox

347

The number of minesweepers that supported the D-Day convoys in World War II.

Strobogrammatic primes on a calculator do not contain the digits 3, 4, or 7.

349

$\pi(349) = \pi(3)^1 + \pi(4)^2 + \pi(9)^3$. Note that 349 is the largest number with this property. [Firoozbakht]

Frame 349 of the original Zapruder slide set (showing the assassination of President John F. Kennedy) is missing from the National Archives.

353

The only three-digit prime such that the sum of each of its digits raised to itself is prime, i.e., $3^{353} + 5^{353} + 3^{353}$ is prime. [Opao]

The sum of the first seventeen palindromic numbers, beginning with 0. [De Geest]

$353^4 = 30^4 + 120^4 + 272^4 + 315^4$. [Norrie]

The smallest multidigit palindromic prime whose digits are all prime. [Gupta]

The sum of the first five primes that are not Chen primes. Note that 353 is a palindromic Chen prime. [Post]

359

The Moscow Puzzles: 359 Mathematical Recreations was authored by a Russian high school teacher named Boris A. Kordemsky and edited by Martin Gardner.

$359 = -1 + 2^3 \cdot 45$. [Kulsha]

In the *Star Trek* fictional universe, the Battle of Wolf 359 was the Federation's first major battle against a group of cyborgs.

367

The Pythagorean Proposition, by early 20th century professor Elisha Scott Loomis, is a collection of 367 proofs of the Pythagorean Theorem.

Pythagoras (c.500 B.C.)

At least 367 people have to be gathered together in order to *ensure* that two of them share a common birthday—but far fewer usually suffices (page 32). [Beedassy]

The largest number whose square (134689) has strictly increasing digits. [Beedassy]

373

Water boils at approximately 373 Kelvin. [Croll]

Describing 373 and repeating the process with each new term produces three more primes, i.e., one 3, one 7, one 3, generates 131713, etc. [Honaker]

The smaller member of a Gridgeman pair (see 181 on page 75). [Patterson]

379

The sum of the seven smallest primes whose last digit is seven. [Gallardo]

"Man o' War" is buried beneath a larger-than-life bronze statue of himself at Kentucky Horse Park, surrounded by the graves of several of his 379 children.

$*54{\cdot}43$. $\vdash :. \alpha, \beta \in 1 . \supset : \alpha \cap \beta = \Lambda . \equiv . \alpha \cup \beta \in 2$

Dem.

$\vdash . *54{\cdot}26 . \supset \vdash :. \alpha = \iota` x . \beta = \iota` y . \supset : \alpha \cup \beta \in 2 . \equiv . x \neq y .$

$[*51{\cdot}231] \qquad \equiv . \iota` x \cap \iota` y = \Lambda .$

$[*13{\cdot}12] \qquad \equiv . \alpha \cap \beta = \Lambda \qquad (1)$

$\vdash . (1) . *11{\cdot}11{\cdot}35 . \supset$

$\vdash :. (\exists x, y) . \alpha = \iota` x . \beta = \iota` y . \supset : \alpha \cup \beta \in 2 . \equiv . \alpha \cap \beta = \Lambda \qquad (2)$

$\vdash . (2) . *11{\cdot}54 . *52{\cdot}1 . \supset \vdash .$ Prop

From this proposition it will follow, when arithmetical addition has been defined, that $1 + 1 = 2$.

Figure 35. A Lemma Towards $1 + 1 = 2$ from *Principia Mathematica*

It was not until page 379 in the first edition of their pivotal *Principia Mathematica* (a 1910 text deriving logic and mathematics from an axiomatic basis) that Whitehead and Russell were able to prove the key lemma (Figure 35).

383

The first multidigit palindromic prime to appear in the decimal expansion of π.

$$\pi = 3.1415926535\ 8979323846\ 264\underbrace{383}279\ 5028841971...$$

[Wu]

The only known multidigit palindromic **Woodall prime**.

The sum of the first three 3-digit palindromic primes. [Vouzaxakis]

389

The smallest conductor of a rank 2 elliptic curve.

397

Conjectured to be the largest prime that can be represented uniquely as the sum of three positive squares ($3^2 + 8^2 + 18^2$). [Noe]

401

A 401(k) is a type of tax deferred retirement plan that allows employees to save and invest for their own retirement (defined in the U.S. Internal Revenue Code 26 U.S.C. §401(k)).

The Magnolia Bakery is located at 401 Bleecker Street in the West Village neighborhood of Manhattan, New York City. The exterior of the bakery is featured in the movie *Prime.*

401 was the hull number of the RMS *Titanic.* [Gronos]

409

The concatenation of the first 409 odd numbers in reverse order is prime. [Gupta]

"FORMULA 409" claims that it "cleans and degreases virtually any hard, nonporous surface; no rinsing required."

A high-powered Chevrolet engine introduced in 1961 measured 409 cubic inches, which was large for its time. It was immortalized by the Beach Boys in the song "409" which starts out, "She's real fine, my 409." [Litman]

419

Dutch microscopy pioneer and naturalist Anton van Leeuwenhoek (1632–1723) ground a total of 419 lenses during his life. [Oakes]

The Nigerian Advance Fee Scheme (also known internationally as "4-1-9" fraud after the section of the Nigerian penal code which addresses fraud schemes) is generally targeted at small and medium sized businesses, as well as charities. [Croll]

421

The first prime formed by the powers of 2 in numerical order from right to left. [Luhn]

In 1937, Lothar Collatz (1910–1990) proposed the following process: take any number larger than one. If it is even, divide it by two. If it is odd, then multiply by three and add one. Stop when you get to 1, otherwise repeat. It is conjectured that this always ends $4 \to 2 \to 1$. For example, $6 \to 3 \to 10 \to 5 \to 16 \to 8 \to 4 \to 2 \to 1$. (Why not try 27 and see why we omitted it from Figure 36?)

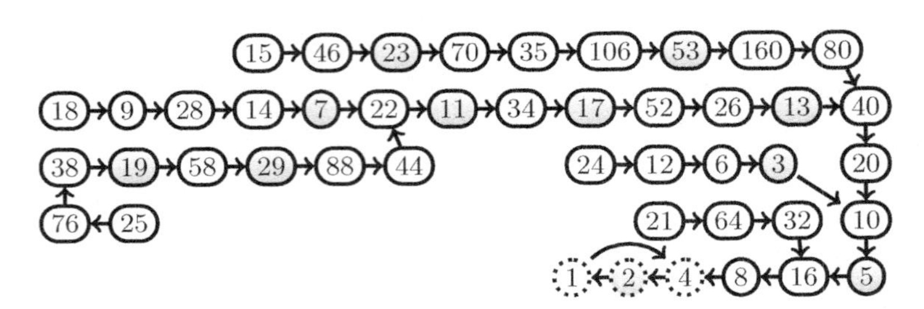

Figure 36. Collatz Conjecture for $n \leq 26$

Jeu du 421 (*quatre-cent-vingt-et-un*) is a French dice game.

The sum of the letters of "prime number" if we use the alphaprime code (page 83). [Zirkle]

431

Polaris (the North Star) is 431 light-years from Earth, according to astrometric measurements of the Hipparcos satellite.

433

Steam locomotive number 433 stands at the trailhead of the Virginia Creeper Trail in Abingdon, Virginia.

The last 433 digits of 433^{433} form a prime number. [Gupta]

Avant-garde composer John Cage's piece titled 4'33" (referred to as "four, thirty-three") entails playing nothing at all for four minutes and thirty-three seconds. (It is probably a coincidence that this is 273 seconds, and absolute zero is very close to $-273\,^{\circ}$C. [Earls]

439

The smallest prime such that another prime is never produced by inserting the same digit between each pair of its digits ($40309, 41319, 42329, \ldots, 49399$, are all composite). [Noll]

443

The first three-digit non-palindromic prime number whose binary equivalent (110111011) is a palindromic prime in base ten. [Hoffman]

449

The smallest prime number whose individual digits are composite perfect squares. [Thoms]

449 is the 87th prime. Note that $4^2 + 4^2 + 9^2 = 8^2 + 7^2$. [Punches]

457

The German submarine known as U-457 was sunk in the Barents Sea by depth charges from the British destroyer HMS *Impulsive* during WWII. [Wynn]

461

Eric Clapton's most famous solo album was called *461 Ocean Boulevard*. This was his comeback record after a long bout with heroin addiction. [Tignor]

$\pi(461)$ is the $(4 \cdot 6 \cdot 1)$th prime. [Firoozbakht]

The prime race $4n - 1$ versus $4n + 1$ is tied at 461. There are no ties in the 4-digit range. [Wilson]

The Knoxville Convention Center contains a 461-seat lecture hall.

463

The smallest multidigit prime such that both the sum of digits and product of digits of its square remain square. [Russo]

463 meters is exactly one-fourth of an international nautical mile. [Honaker]

The entry to the *Duomo* in Florence, from the *Porta della Mandorla* on the north side of the *Cattedrale di S. Maria del Fiore*, contains exactly 463 steps.

467

The smallest prime p in which the concatenation of p with the next prime remains prime throughout two steps of the same procedure. For example, 467 concatenated with the next prime (479) gives the prime 467479, and 467479 concatenated with the next prime (467491) gives the prime 467479467491. 467 is also the smallest whose successive concatenations remain prime throughout three steps. [De Geest and Post]

479

The largest known prime number that cannot be represented as the sum of less than nineteen fourth powers.

487

A prime p such that the decimal fraction $\frac{1}{p}$ has the same period length as $\frac{1}{p^2}$. [Richter]

487 is the smallest prime p such that p and p^3 have the same sum of digits. [Honaker]

Fermat claimed (correctly) that a number is the sum of three squares unless it is of the form $4^n(8m+7)$, with $n, m \geq 0$.

491

The smallest irregular prime having an irregularity index of 3. The irregularity index of a prime p is the number of times that p divides the Bernoulli numbers B_{2n} for $1 < 2n < p-1$. **Regular primes** have an irregularity index of zero. [Noe]

499

The decimal expansion of 499^{499} ends with the digits 499499. Often p^p ends with the digits of p, but this is the only known case for which it ends with the digits of p twice.

The distance light travels the vacuum of space in 499 seconds is approximately the mean distance between the centers of the Earth and Sun (an Astronomical Unit).

$497 + 2$ is the reversal of $497 \cdot 2$.

A knight's tour is a numbered tour of a knight over an otherwise empty chessboard visiting each square once only. A queen placed on the start of a tour discovered by George Jelliss can attack all of the odd primes, and every odd number attacked by the queen is a prime. It is also a more restricted version known as a re-entrant tour, in

04	**07**	10	**23**	64	**19**	12	15
09	24	**05**	18	**11**	14	63	20
06	**03**	08	♕	22	**61**	16	**13**
25	30	**37**	60	**17**	50	21	62
36	**59**	02	**29**	42	**53**	46	51
31	26	33	38	49	40	**43**	54
58	35	28	**41**	56	45	52	**47**
27	32	57	34	39	48	55	44

Figure 37. A Knight's Tour

which the knight, on its 64th move, could arrive back at its starting square. The sum of the odd primes in the tour is 499 (Figure 37).

503

The smallest prime that is the sum of cubes of the first n primes ($2^3 + 3^3 + 5^3 + 7^3$). [Honaker]

Theatre503 is a performing arts venue in London that specializes in new work.

509

The sum of three consecutive squares ($12^2 + 13^2 + 14^2$). [Schlesinger]

The 509th Composite Group (an air combat unit of the U.S. Air Force) fulfilled its mission when the *Enola Gay* piloted by Colonel Tibbets dropped the first atomic bomb on Hiroshima. [McCranie]

521

The smallest prime whose reversal is a cube. [Honaker]

The first successful identification of a Mersenne prime by means of electronic digital computer was achieved in the year 1952, using the U.S. National Bureau of Standards Western Automatic Computer (SWAC) at the Institute for Numerical Analysis at the University of California, Los Angeles. It was $M(521)$.

$M(521)$ can be written as $512 \cdot 2^{512} - 1$. Therefore, it is a Woodall prime as well as a Mersenne prime. [Dobb]

521 is the smallest Mersenne prime exponent that exceeds the sum of all smaller ones. [Terr]

The lesser prime of the only pair of twin primes less than one thousand for which their cubes, when reversed, form primes. [Trotter]

If p is prime, then it divides the pth term of the Perrin sequence: $0, 2, 3, 2, 5, 5, 7, 10, 12, 17, \ldots$ (each term is the sum of the two terms preceding *the term before it*). Often, if $n > 1$ divides the nth term, then n is prime. The first of infinitely many exceptions to this rule is the square of 521. There are only 17 such composites less than 10^9.

523

Together with 541 form the smallest two consecutive primes, such that the sums of the digits are equal. [Smart]

The only three-digit prime containing all three of the first three prime digits. [Patterson]

541

The numerical weight of the name *Israel.* Note that 541, the 100th prime number, is the 10th hexagonal star number. (The figure on the right illustrates the first few of these numbers.) [McGough]

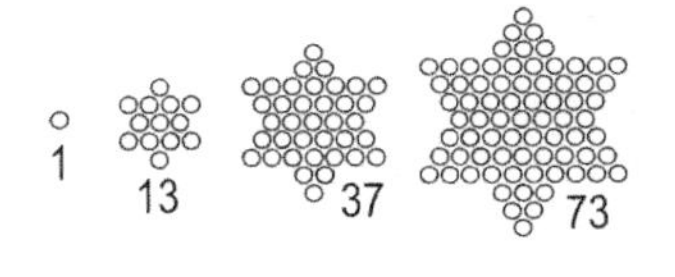

547

The first edition of Crandall and Pomerance's excellent *Prime Numbers: A Computational Perspective* (2002) contained exactly 547 pages.

557

The first and only prime $p < 10^5$, where $(p-1)\# + 2 \equiv 0 \pmod{p}$. [Luhn]

563

Wilson's theorem tells us that $p > 1$ is prime if and only if it divides $(p-1)! + 1$. Wilson primes are those for which $(p-1)! + 1$ is divisible by p^2. The theorem was actually explained by the Iraqi scientist Ibn al-Haytham (also known as Alhazen) around A.D. 1000, but was named after John Wilson (a student of Edward Waring) who stated it in the 18th century. The largest of the three known Wilson primes is 563.

Reggie Jackson hit 563 home runs in his professional baseball career. He led his teams to five world championships and eleven division titles.

569

The number 569 and its twin prime share an interesting property: both fib(569) and fib(569 + 2) are primes as well. [Dobb]

571

NBC News reported in June 2001 that there are 571 stoplights along U.S. Route 66.

Commissioned in 1954, the USS *Nautilus* (SSN-571) was the world's first nuclear-powered submarine. [Mostow]

577

The first three digits in the decimal expansion of the Euler-Mascheroni constant $\gamma = 0.5772156649....$ G. H. Hardy (1877–1947) is said to have offered to give up his Savilian Chair at Oxford to anyone who proved that it was not a rational number (i.e., not a ratio a two integers). No one has yet to do so.

$$\gamma = \lim_{n \to \infty} \left[\left(1 + \frac{1}{2} + \frac{1}{3} + \frac{1}{4} + \cdots + \frac{1}{n} \right) - \log n \right]$$

[Kulsha]

577 is one of only two numbers with squares of the form AABCBC. Can you find the other? [Axoy]

587

Charleston endured 587 days of constant shelling by Federal forces, both on land to the south and at sea near the mouth of the harbor during the U.S. Civil War. [Cambell]

Nickel Plate No. 587 (at the Indiana Transportation Museum) is perhaps the best remaining example of a United States Railroad Administration light Mikado steam locomotive.

Let $a(1) = 7$ and $a(n) = a(n-1) + \gcd(n, a(n-1))$, for $n > 1$. Here "gcd" means the greatest common divisor. Recently, Rutgers graduate student Eric Rowland proved that $a(n) - a(n-1)$ is either 1 or a prime! The differences begin 1, 1, 1, 5, 3, 1, 1, 1, 1, 11, 3, 1, 1, 1, 1, 1, 1, 1, 1, 1, 1, 23, ... It is not known if all odd primes occur in this list. Jonathan Sondow found that 587 is the smallest odd prime that does not appear in the first 10000 prime terms.

593

$593 = 9^2 + 2^9$. [Astle]

The song "Re-Make/Re-Model" from Roxy Music's eponymous debut album contains the esoteric chorus "CPL 593H" (referring to a British license plate).

599

The number in the street address of the Administrative building at Saint-Prime (599, rue Principale, Saint-Prime, Québec, Canada). [Blanchette]

601

The PowerPC 601 Microprocessor is a highly integrated single-chip processor that combines a powerful RISC architecture, a superscalar machine organization, and a versatile high-performance bus interface. Superscalar refers to microprocessor architectures that enable more than one instruction to be executed per clock cycle.

00601 is the smallest assigned 5-digit ZIP Code that is prime. It belongs to Adjuntas, PR. Note that Adjuntas is also known as "La Suiza de Puerto Rico" (the Switzerland of Puerto Rico) because of its low temperatures. [Wolfe]

The aliquot sequence that starts with 446580 ends (4736 iterations later) with 601. [Creyaufmüeller]

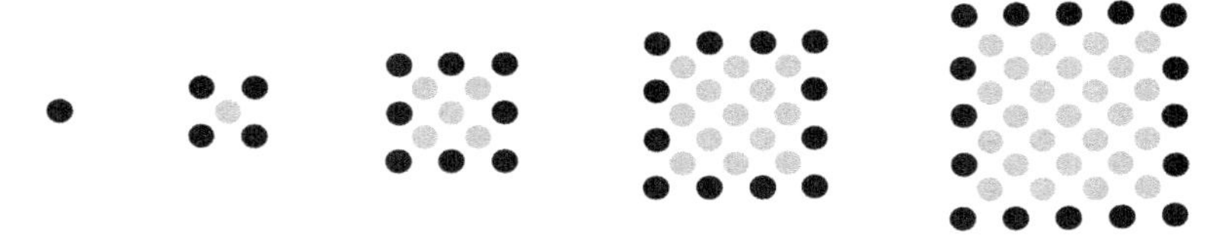

Figure 38. Centered Square Numbers: 1, 5, 13, 25, 41, ...

607

If displayed on a calculator upside down, 607 spells Log (the abbreviation for logarithm). [Lipps]

One *stadium*, about 607 feet, was the length for the course in the ancient Greek Olympic Games at Olympia. (The name for the length of the race eventually became the name for the place in which it was run, i.e., a stadium.) [Motz]

613

NGC 613 is a barred spiral galaxy in the southern constellation Sculptor. The sum of digits of 613 approximates its visual magnitude.

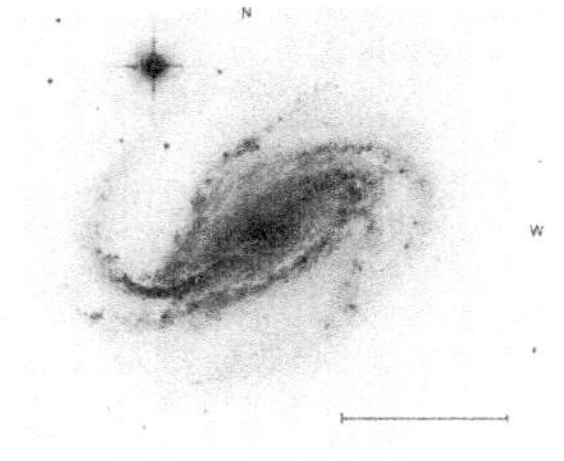

Galaxy NGC 613

613 is a mathematical enigma in the bewildering story *Number of the End* by Jason Earls. "Bring the first digit back to get 136, it's triangular. Now bring the first digit of that back to get 361, it's a square."

There are exactly 613 precepts (or laws) within the five books of Moses. [Aaron]

The 18th centered square number (similar to squares; Figure 38). Note that $18 = 6 \cdot 1 \cdot 3$. [Post]

617

The largest multidigit prime that is exactly half of a number formed with distinct consecutive digits. [Dickman]

The largest of the **RSA numbers** has a length of 2048 bits (617 decimal digits). There was once a $200,000 prize offered for its factorization, but is now withdrawn. Anyway, RSA-2048 is unlikely to be factored anytime soon.

619

6! − 5! + 4! − 3! + 2! − 1!. [Guy]

According to Urantia, a religious sect headquartered in Chicago, we live on a planet in a system that includes 619 flawed but evolving worlds.

The smallest strobogrammatic prime that is not a palindrome. [Punches]

The first person to be converted to Islam (Khadija, Prophet Muhammad's first wife) died in 619.

631

The reverse concatenation of the first three triangular numbers. [Gupta]

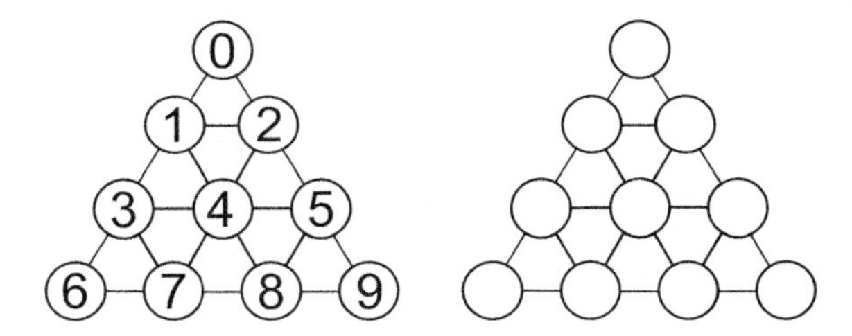

Figure 39. The Tetractys Puzzle (631)

The tetractys (pronounced "tet-trak'tis") is a triangular figure consisting of ten vertices arranged in four rows: one, two, three, and four dots in each row. It was a mystical symbol to the Pythagoreans, who lived during the 6th century B.C. There are fifteen primes in Figure 39 (reading forwards or backwards along the indicated lines), the largest of which is 631. Can you rearrange these digits and achieve two dozen primes? (The first such solution was found by an individual serving time in a juvenile detention center.)

The largest known difference between consecutive Ulam numbers (page 31): 332250401 and 332251032. [Knuth]

641

Replacing each digit d of 641 with d copies of the digit d produces another prime, i.e., 641 becomes 66666644441, which is also prime. If we apply the same transform using 66666644441, yet another prime is formed. [Honaker]

Pulsars (rapidly spinning neutron stars) have been observed to spin up to 641 times per second.

In 1732, Euler found that $2^{2^5} + 1 = 4294967297$, the first composite Fermat number, is divisible by 641.

The 193-digit integer $2^{641} - 1$ was the smallest Mersenne number that had not been completely factored prior to the 21st century. [Cerias]

The Great Library of Alexandria, home to such mathematicians as Euclid, Archimedes, Eratosthenes, Apollonius, and Pappus, had disappeared before the arrival of the Muslim Arab armies in A.D. 641.

Archimedes on the Fields Medal

The Ford 641 Workmaster is an antique farm tractor. They have been known to crop up at the annual Lee County Tobacco & Fall Festival in Pennington Gap, Virginia.

Room 641A is an alleged intercept facility operated by AT&T in San Francisco for the U.S. National Security Agency. [McCranie]

643

The largest prime factor of 123456. Note that $643(64 \cdot 3) = 123456$. [De Montagu]

In baseball, a double play which begins with the shortstop and is then thrown to the second baseman is called a 6-4-3 double play. [Patterson]

$643 = \frac{3^8+8^3}{3+8}$. [Trotter]

647

The smallest prime that may be written as $2a^2 - 1$, and also as $3b^3 - 1$. [Hartley]

A man described as looking like a plump Moses asked the following questions during a 1987 joint American Mathematical Society/Mathematical Association of America conference in San Antonio, Texas: "What about 647? Is it prime?" (from *The Man Who Loved Only Numbers* by Paul Hoffman)

653

The first three-digit prime number to occur in the decimal expansion of π (see page 90).

659

The first prime for which we do not include a curio in this edition.

661

Lord Kelvin, one of Britain's great physicists, published 661 papers on a wide range of scientific subjects. [Rosen]

661 is the start of a record-breaking twin prime gap.

673

The Louvre Pyramid is made up of 673 glass lozenges and triangles, not counting the doors.

Chlorophyll *a* fluoresces at 673 nanometers.

677

The sum of three consecutive squares $(14^2 + 15^2 + 16^2)$. [Schlesinger]

691

G. N. Watson proved that Ramanujan's tau function $\tau(n)$ is divisible by 691 for almost all positive integers (see page 228). [Terr]

The first irregular prime to appear in the numerator of a Bernoulli number.

The first prime Lychrel number (a number that does not form a palindrome by repeatedly applying the reverse-then-add process). The first Lychrel number is the reversal of 691.

The smallest prime that can be written as the sum of thirteen consecutive primes. Recall that 691 is thirteen squared, turned upside down. [Post]

701

701, 7001, 70001, and 700001 are each prime. [Brown]

The number of UFO cases that Project Blue Book left classified as "unknown." [McCranie]

There are currently 701 types of pure breed dogs. [BBC]

701 can be written as $5^4 + 4^3 + 3^2 + 2^1 + 1^0$. [Kulsha]

The IBM 701 was the company's first production-line electronic digital computer. (It was announced to the public in 1952.)

709

709 ones followed by 709 is prime. [Wilson]

The smallest prime whose cube is the sum of three prime cubes: 709^3 is $193^3 + 461^3 + 631^3$. [Rivera]

719

A number that can be expressed as both a **factorial prime** $(6! - 1)$ and as an abundant number less one. (A number is abundant if it is smaller than the sum of its proper divisors, e.g., $12 < 1 + 2 + 3 + 4 + 6$; the classification of numbers into deficient, perfect, and abundant was first made by Nicomachus of Gerasa in his *Introductio Arithmetica*, circa A.D. 100.) [Patterson]

The smallest factorial prime of the form $n! - 1$ such that $n - 1$ is a factorial prime of the same form. [Capelle]

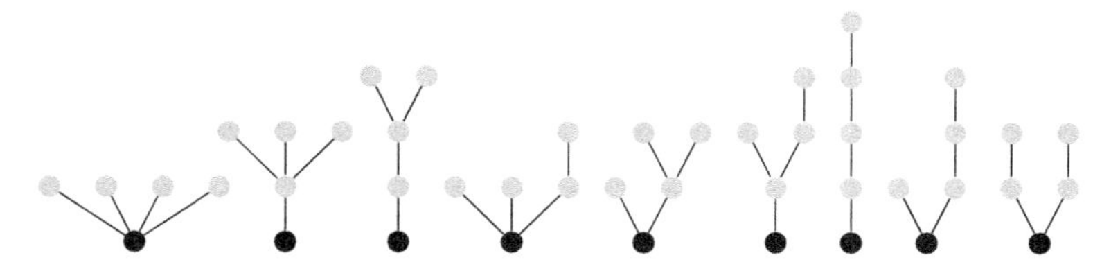

Figure 40. Example: There are Nine Rooted Trees with Five Nodes

The number of rooted trees (Figure 40) with 10 nodes and also the number of ways of arranging 9 nonoverlapping circles. [Beedassy]

The first consecutive primes up to 719 will form a **Smarandache-Wellin prime** if concatenated in increasing order.

The hour and second hands of an analog clock align exactly 719 times every twelve hours. [Rosulek]

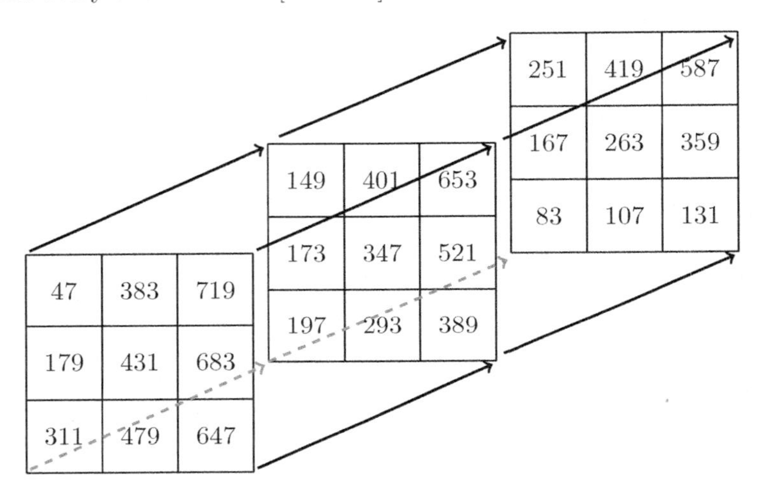

Figure 41. The Smallest 3-by-3-by-3 Balog Cube of Primes

Antal Balog proved there are infinitely many 3-by-3-by-3 cubes of distinct primes, where each line of 3 primes parallel to any edge is an arithmetic progression. The largest prime in the smallest 3-by-3-by-3 Balog cube is 719 (Figure 41). Green and Tao showed there are such cubes for all larger dimensions using n-by-n-by ... by-n arrays. [Granville]

727

The smallest odd prime that can be represented as the sum of a cube and its reversal $(512 + 215)$. [Gupta]

The Boeing 727 became the first three-engine jet built for commercial service.

The first prime whose square (528529) can be represented as the concatenation of two consecutive numbers. [De Geest]

$727 = 1! + (1 + 2)! + (1 + 2 + 3)!$. [Van Doorn]

733

733 (MHz) was the largest prime "chip" among 20th century processors. [Kulsha]

$733 = 7 + 3! + 3!!$ is prime. [Meyrignac]

$733 = 1^7 + 2^6 + 3^5 + 4^4 + 5^3 + 6^2 + 7^1 + 8^0$. [Vrba]

739

The first Human Genome Project online download contained a 739 megabyte file.

The smallest prime number that is the sum of distinct primes beginning and ending with the digit three ($739 = 3 + 353 + 383$). [Capelle]

743

The digits of 743 appear as exponents in $648^7 + 139968^4 = 7558272^3$.

The Sun is a gaseous mass about 743 times heavier than the total planetary mass. [Bertotti and Farinella]

751

The largest prime that cannot be expressed as the sum of five or fewer squared composite numbers.

Chicago psychologist Doctor Robert Hartley's suite number on the TV series *The Bob Newhart Show* is 715, but has appeared as 751 due to the rearrangement of digits by set designers.

The number built with "expanded notation" in the classic book *MATHEMATICS* by David Bergamini and the Editors of LIFE (1963, p. 194), revealing to primary-graders how large numbers are constructed of hundreds, tens, and units.

757

For years the Boeing 757 had the lowest operating cost per seat-mile of any single-aisle jetliner in its class and a lower cost per trip than any twin-aisle airplane.

The California lottery announces results on TV at precisely 7:57 P.M. [Haga]

761

A sequence of six 9's (known as the Feynman Point) begins immediately after the 761st decimal place of π. Nobel prize-winning physicist Richard Feynman expressed a wish to memorize the digits of π as far as that point so that when reciting them, he would be able to end with "...nine, nine, nine, nine, nine, nine, and so on."

773

The largest three-digit **unholey prime**.

The only three-digit **iccanobiF prime**.

The smallest prime such that adding a preceding or trailing digit will always result in a composite if we omit the case where the leading digit is zero. [Noll]

787

The six integers following 787 are divisible by the first six primes, respectively.

The smallest prime that can be represented as sum of a prime and its reversal in two different ways. [Gupta]

797

The largest palindromic two-sided prime (both right and left-truncatable). [Gupta]

809

Formerly the area code for the entire Caribbean. According to the National Fraud Information Center, it has been associated with long distance phone scams.

The hotel room number in the 1952 film *Don't Bother to Knock*, starring Richard Widmark and Marilyn Monroe. Marilyn's character (Nell Forbes) is a lonely blonde who has recently been released from a mental institution. [McCranie]

The only three-digit circular-digit prime.

Table 3. Minimal Primes in Small Bases

base	**minimal primes** (written in that base)
2	10, 11
3	2, 10, 111
4	2, 3, 11
5	2, 3, 10, 111, 401, 414, 14444, 44441
6	2, 3, 5, 11, 4401, 4441, 40041
7	2, 3, 5, 10, 14, 16, 41, 61, 11111
8	2, 3, 5, 7, 401, 661, 6441, 444641, 444444441
9	2, 3, 5, 7, 14, 18, 41, 81, 601, 661, 1011, 1101
10	(see 66600049 on page 194)

811

The largest minimal prime in base nine (Table 3). This number is 1101 in base nine, i.e., $811 = 1 \cdot 9^3 + 1 \cdot 9^2 + 0 \cdot 9 + 1$. [Rupinski]

821

The smallest prime whose reversal is a seventh power ($128 = 2^7$). [Gupta]

The smallest prime of the first prime quadruple for which the sums of the cubes of the digits of the four primes (821, 823, 827, 829) are primes themselves (521, 547, 863, 1249). [Trotter]

823

Victor Hugo's *Les Misérables* contains one of the longest sentences (823 words without a period) in the French language.

827

The smallest prime formed from the concatenation of two consecutive cubes. [Gupta]

The Unabridged Running Press Edition of the American classic *Gray's Anatomy* contains 827 illustrations.

829

The smallest prime factor of the smallest composite 2 and 3-SPRP (strong **probable prime**).

839

$(839-8)^8+(839-3)^3+(839-9)^9$ is prime. [Opao]

853

853 divides the sum of the first 853 prime numbers (2615298).

The Transamerica Pyramid has a structural height of 853 feet. It is the tallest skyscraper in San Francisco, California.

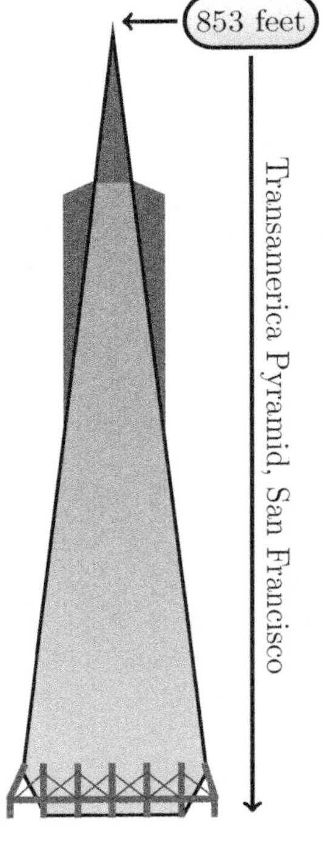

859

The largest prime that cannot be the central number in a 3-by-3 prime magic square. [Rivera]

Théorie des Nombres by Adrien-Marie Legendre (1752–1833) contained 859 large quarto pages and was the first text to *record* the connection between the **prime counting function** $\pi(x)$ and the logarithm function. Legendre stated $\pi(x) \sim \frac{x}{\log x - 1.08366}$ when x goes to infinity, though $\frac{x}{\log x - 1}$ is a better choice in the long run.

881

Hill 881 was the site of fierce and bloody fighting between soldiers of the North Vietnamese Army and United States Marines during the Vietnam War. The number refers to the elevation of the hill in meters.

883

$8^n+8^n+3^n$ is prime for $n=0$ to 4. [Neofytou]

Penélope Cruz has 883 tattooed on her right ankle. According to numerologist Christine DeLorey, author of the hit book *Life Cycles*, eight and three are the actress's main numbers in life. Eight stands for money, power, and success, while three stands for skill in communications and creativity. The second eight is used to strengthen the power of the first eight.

907

Conjectured to be the largest prime that can be represented uniquely as the sum of three distinct positive squares ($3^2 + 13^2 + 27^2$). [Noe]

907, 9907, 99907, 999907, and 9999907 are all primes; however, 99999907 is composite.

911

On the first anniversary of the 9/11 terrorist attacks in New York City, the numbers that popped up for the New York Lottery were 9-1-1. [MSN]

9-1-1 is the prime phone number in the United States to dial in case of emergency.

The Madrid Attack came 911 days after 9/11.

Consider a Fibonacci-like sequence starting with 9, 11. The 11th term will be 911.

919

The smaller number in the smallest pair of prime numbers that are mutually the sums of (the same) powers of each other's digits, i.e., $919 = 1^3 + 4^3 + 5^3 + 9^3$ and $1459 = 9^3 + 1^3 + 9^3$. [Hartley]

A palindromic prime containing nineteen letters in its English name "nine hundred nineteen." Note that $9 + 1 + 9$ equals nineteen and you get nineteen in either direction (left or right) from the center. [Post]

The smallest number that cannot be added to a nonzero palindrome such that the sum is also palindromic. [De Geest]

The largest known palindromic prime for which the next prime is also palindromic. [Beedassy]

Modern studies have shown that the earliest known version of the beast number (666) may have been 919 turned upside down.

929

The number of chapters in the Old Testament of the Protestant Bible (e.g., KJV). [Agard]

$\pi(929) = 9^2 - 2^2 + 9^2$.

The smallest palindromic prime whose cube can be expressed as the sum of three odd cubes: $929^3 = 69^3 + 447^3 + 893^3$. [Rivera]

$9^{29} + 92^9$ is prime. [Vrba]

937

The number of Germans that set a mass yodeling record in 2003. [Dobb]

941

The smallest prime formed from the reverse concatenation of three consecutive squares. [Gupta]

971

The only prime in the alphametic on the right. [Madachy]

```
   UK
  USA
+ USSR
------
ABOMB
```

977

In the movie *Harvey*, Dr. Sanderson prescribes "Formula 977" to Elwood P. Dowd (James Stewart). [La Haye]

If we use the alphaprime code (page 83), the sum of the letters of "nine hundred and seventy-seven" is a prime adjacent to 977. It is left for the prime curiologist to find if the adjacent prime occurs before or after 977. [Hunnell]

The first prime that results in a cube number if added to the sum of its digits. [Bopardikar]

983

The number of words in Theodore Roosevelt's Inaugural Speech. [Blanchette]

Enoch Haga, author of *Exploring Prime Numbers on Your PC and the Internet*, lived at 983 Venus Way in Livermore, California, at the time of its publication.

991

991, 99991, and 9999991 are each prime. [Avrutin]

997

The smallest prime of the form $10^n - n$. [Luhn]

1009

The smallest prime or emirp greater than a thousand. [Gevisier]

The smallest prime that can be expressed in the form $x^2 + ny^2$ for all values of n from 1 to 10. [Ashbacher]

The smallest number which is the sum of three distinct positive cubes in more than one way: $1^3 + 2^3 + 10^3$ and $4^3 + 6^3 + 9^3$. [Beedassy]

MIX's model number in Donald Knuth's monograph, *The Art of Computer Programming*. ("MIX" equals 1009 in Roman numerals.) [Nowacki]

1013

1013 Productions is the company owned and managed by Chris Carter, who created *The X-Files*. [Blanchette]

The average sea-level pressure on Earth is about 1013 millibars. [Webb]

In Marcus du Sautoy's book *The Music of the Primes*, while discussing **Bertrand's postulate**, he states that there are actually quite a lot of primes between 1009 and 2018, the first being 1013.

A free neutron decays with a half-life of about 1013 seconds into a proton, an electron, and an antineutrino. [Hughes]

1019

The 20th century musical *Damn Yankees* had a successful run of 1019 performances. [Williams]

1031

The final bid on eBay for a possible Erdős number of 5 was $1031 (April 30, 2004). The high bidder refused to pay, saying he had bid only to "stop the mockery," adding "papers have to be worked and earned, not sold, auctioned or bought." (Those who coauthored an article with Erdős have an Erdős number of 1; those who coauthored with those have an Erdős number of 2; and so on.) Almost 90000 people have an Erdős number of 5. [McCranie]

Erdős (1913–1996)

The number of ones that form the fifth and largest known repunit prime. It was first proven prime in 1985. (The next few that could possibly be prime are those with 49081, 86453, 109297, and 270343 digits.)

$1031 = 2 \cdot 3 \cdot 5 + 7 \cdot 11 \cdot 13$. [Capelle]

Halloween falls on "ten thirty-one" (10/31) each year. It is a cross-quarter date, approximately midway between an equinox and a solstice.

1033

$8^1 + 8^0 + 8^3 + 8^3 = 1033$. "My math students 'ate' this up!"

1039

The center prime number in the smallest possible 3-by-3 prime magic square consisting of primes in arithmetic progression (but not consecutive).

"1,039/Smoothed Out Slappy Hours" is a collection of early recordings by Green Day (an American rock band formed in 1987). [Rupinski]

1049

The smallest prime containing all of the square digits exactly once. [Gupta]

Phil Appleby of the United Kingdom achieved the highest competitive game score of 1049 in *Scrabble* on June 25, 1989 (according to the *Guinness Book of World Records*). [Patterson]

1051

C. J. Mozzochi utilized 1051 to show that if we let p_n be the nth prime, then there exists a constant K such that $p_{n+1} - p_n < Kp_n^{\frac{1051}{1920}}$. [Post]

1061

The smallest **bemirp** (or bi-directional emirp).

A four-digit prime that equals the number of primes with four digits. [Russo]

The Normans conquered Messina in 1061.

1069

The smallest four-digit emirp with distinct digits. [Capelle]

1091

$n^6 + 1091$ is composite for all values of n from 1 to 3905. [Shanks]

1093

Thomas A. Edison held 1093 successful U.S. patent applications. [Dobb]

Edison (1847–1931)

The smallest **Wieferich prime**: 1093^2 divides $2^{1093-1} - 1$.

1103

Ramanujan produced an excellent approximation to $2\pi\sqrt{2}$ by dividing 99^2 by 1103. [Croll]

$1103^2 = 1216609$ and $3011^2 = 9066121$. [Kulsha]

The smallest reflectable balanced prime. [Abramowitz]

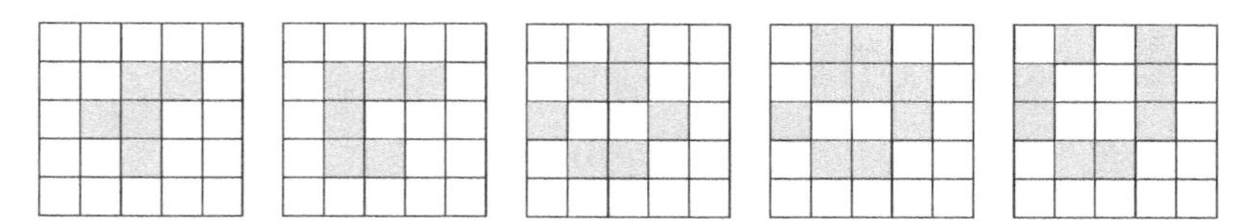

Table 4. Five Generations in Conway's Game of Life

John Conway's game of life starts with cells that are "alive" (marked or shaded) in an infinite array. Each subsequent "generation" (iteration): a living cell stays alive if it has either two or three living neighbors; and a dead cell springs to life if it has exactly three live neighbors. The R-pentomino and its next four generations are shown in Table 4. The pattern that this pentomino generates does not stabilize until the 1103rd generation. (There is even an initial arrangement of cells that will generate the prime numbers!)

1117

The concatenated prime factors of $\pi(1117)$. [Lewis]

The famous Japanese mathematician Yutaka Taniyama (1927–1958) committed suicide on November the seventeenth (11/17). Among his achievements was a key step toward the proof of Fermat's Last Theorem.

1123

11/23 (November 23) is the only date containing the concatenation of two 2-digit primes that can be read as a prime number in either "month-day" or "day-month" format, i.e., 11/23 or 23/11. Note that Kinro Kansha no Hi (Labor-Thanksgiving Day) occurs on this date in Japan. [Kik]

A prime formed by concatenating the first four Fibonacci numbers in sequence. [Gupta]

All the residences in Primes Lane in Holton, Suffolk, England, have a prime postcode. Primes Lane leads southwards into the B1123 main road. [Croll]

1129

The smaller member in the least set of "blackjack primes," i.e., primes separated by exactly 21 consecutive composite numbers.

1151

In the 13th century, Persian mathematician Kamal al-Din Abu'l Hasan Muhammad ibn al-Hasan al-Farisi used a number of lemmas including an application of the **Sieve of Eratosthenes** to show that 1151 is prime.

1193

The start of the smallest sequence of ten consecutive emirps. [Rivera]

1213

The smallest four-digit prime (emirp) containing consecutive numbers (12 and 13). [Das]

1229

There are exactly 1229 prime numbers less than ten thousand. [Haga]

1231

The smallest prime that can be represented as the sum of a prime and its reversal in three different ways. [Gupta]

The last "prime day" of the year is 12/31. [Hallyburton]

1237

The difference between 1237^2 and $(1237 - 5)^2$ might surprise you.

1259

Twelve fifty-nine (12:59) is the largest "prime time" of day on a 12-hour clock in hours and minutes.

The first prime formed from the leading digits of the decimal expansion of the Delian constant ($\sqrt[3]{2} = 1.259...$). See page 167. [Gupta]

1291

The lesser prime in the smallest set of five consecutive primes whose sum of digits are another set of distinct primes. [Gupta]

1297

The smallest prime whose reversal is a Fibonacci number squared (7921 is 89^2). [Gupta]

1301

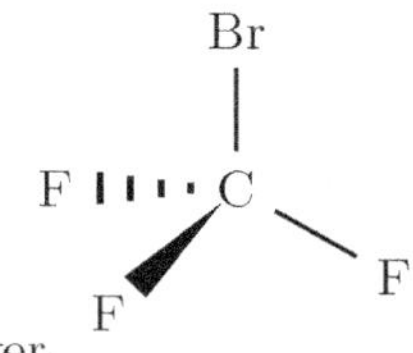

Bromotrifluoromethane, also known by the trade name Halon 1301, is a colorless, odorless, and non-toxic gas which extinguishes a fire by chemically reacting with the combustion process. It is now known to be environmentally damaging, particularly toward the Earth's protective ozone layer.

1319

The buildings inside the Cheyenne Mountain Complex (outside Colorado Springs, Colorado) which house the North American Aerospace Defense Command (NORAD) are supported on 1319 springs to protect the delicate electronic equipment inside the mountain from shock. [McCranie]

1327

The number of named openings and variations listed in the second edition of *The Oxford Companion to Chess* by Hooper and Whyld.

1361

The sum of squares of the first seven Lucas numbers. [Gallardo]

1423

Joy of Thinking: The Beauty and Power of Classical Mathematical Ideas is The Teaching Company's course number 1423.

1427

The smallest four-digit prime such that the following form is also prime: $142^7 + 14^{27} + 1^{427}$. [Patterson]

1429

The sum of two famous baseball records: the number of home runs hit by Babe Ruth (714), and the number of the home runs hit by Hank Aaron to break the Babe's record (715). The pair of numbers 714 and 715 is called a Ruth-Aaron pair because the sums of the prime factors of these consecutive integers are equal. The number $714 \cdot 715$ is also the product of the first seven primes (i.e., 7-**primorial**). [Trotter]

The concatenation of the first 1429 prime numbers is the largest known Smarandache-Wellin prime.

1447

The smallest **straight-digit prime** that contains each of the straight digits (1, 4, and 7). [Gupta]

The smallest prime that is pandigital in Roman numerals, i.e., using each of the symbols I, V, X, L, C, D, and M at least once. [Post]

1459

The only four-digit prime number such that the sum of the first and second digits is the third digit, the sum of the second and third digits is the fourth digit, and the sum of the third and fourth digits is the first two digits. [Rupinski]

1483

The Petronas Towers in Kuala Lumpur, Malaysia, is the tallest (1483 feet) twin towers in the world, and it lays claim to being the world's tallest high-rise of the 20th century.

1493

The year that Christopher Columbus returned to Spain after his first New World voyage.

The largest known Stern prime, i.e., it is not the sum of a smaller prime and twice the square of a nonzero integer. These are named after Jewish mathematician Moritz Abraham Stern (1807–1894). (Altogether, only eight are known: 2, 3, 17, 137, 227, 977, 1187, and 1493.)

1499

An emirp that remains prime if any digit is deleted. [Post]

1511

It is alleged, yet unproven, that a monk was killed by a meteorite in Cremona, Italy, in 1511. In algebraic geometry, the problem of describing the Cremona group of spaces in three dimensions and higher has not been settled either. [Sullins]

1531

The smallest prime number that is a sum of distinct primes beginning and ending with the digit seven ($7 + 727 + 797$). [Capelle]

1549

The smallest multidigit number that is not the sum of a prime and a power.

1553

The number of ways of rearranging nine people seated around a circular table, so that no one sits next to any of his neighbors from the previous arrangement. [Post]

1559

United Nations Security Council Resolution 1559 was an attempt to discourage Syrian meddling in Lebanon.

1579

The hull number of *Cape Keltic* in the Hallmark Hall of Fame movie *Calm at Sunset.* [Haga]

1597

The largest known Fibonacci emirp.

There exist positive values of n such that $\sqrt{1597n^2+1}$ is an integer, but you'll need more than a hand-held calculator to find even the smallest solution.

Jacopo Peri (1561–1633) wrote the first work to be called an opera today, *Dafne* (circa 1597, now lost). Note that 1597 is the arithmetic mean between 1561 and 1633.

Fermat (1601–1665)

1601

Anti-nuclear vigils are held at 1601 Pennsylvania Avenue (Lafayette Park) in Washington, D.C.

Pierre de Fermat, "The Prince of Amateurs," was born in 1601. Did he know that 1601 to the power 1601 ends with 1601? [Capelle]

1607

The first permanent English settlement in the New World took place in 1607 on Jamestown Island, Virginia.

1609

The house at 1609 16th St. NW in Washington, D.C. (H. Cornell Wilson House) was the home/office of the Einstein-like scientist

Dr. Barnhardt in the classic movie *The Day the Earth Stood Still.* [McCranie]

1619

A prime whose cube can be expressed as the sum of three prime squares in two different ways, i.e., 1619^3 is $3^2 + 24967^2 + 60169^2$ and $3^2 + 28163^2 + 58741^2$. [Rivera]

1627

The start of the first occurrence of three consecutive primes ending with the digit seven (1627, 1637, 1657). [Murthy]

1637

René Descartes introduced the mathematical terms real and imaginary in his work *La Géométrie* in 1637. [Gevisier]

Descartes (1596–1650)

1667

Vega was bleating out prime numbers in the 1667 megahertz hydroxyl line in *Contact* (Carl Sagan's only science fiction novel).

The year in which the English Parliament passed a law against fining or imprisoning jurors for returning the "wrong" verdict. [NYU Law Review]

1699

The largest prime that cannot be expressed as the sum of at most five distinct squared composite numbers.

1709

In the middle of 1709 insert 57 and you get 175709, which is prime. Put two 57's in the middle of 1709 and you get 17575709, which is another prime. Continue this process for a sequence of eight consecutive prime terms.

1741

The smallest prime p such that p^9 is equal to the sum of 9 consecutive primes. [Rivera]

1747

The number of digits in the iterated factorial 3!!!. [Hartley]

1759

The last time Halley's Comet passed through its perihelion in a prime year was 1759.

The number of reversible primes below one hundred thousand. [Beedassy]

1777

Carl Friedrich Gauss, "The Prince of Mathematicians," was born in 1777.

Gauss (1777–1855)

1783

Leonhard Euler, the most prolific mathematician in history, died in 1783. [Beisel]

1789

The year the first President of the United States (George Washington) took office.

The French mathematician Augustin-Louis Cauchy (1789–1857) was born with the French Revolution. [Luhn]

Cauchy (1789–1857)

The year Baptist minister Elijah Craig established a still in Georgetown, Kentucky, and began producing America's first bourbon whiskey from a base of corn.

1801

The first full and correct proof of the Fundamental Theorem of Arithmetic was first published by Gauss in his classic work *Disquisitiones Arithmeticæ* (1801). The theorem states that every natural number greater than 1 can be written as a product of primes in exactly one way (apart from rearrangement).

Contemporary painter Michael Eastman employs numbers, letters, and a William Morris pattern as abstract elements in the oil on canvas painting *1801: 14 Prime Numbers.*

The smallest prime consisting of all of the cube digits (i.e., 0, 1, and 8) at least once. [Gupta]

1811

The product of 1811 and the next two consecutive primes results in a concatenation of two consecutive primes in descending order.

$$1811 \cdot 1823 \cdot 1831 = \underbrace{60449}\underbrace{60443}$$

[De Geest]

The year self-educated Peter Barlow published *An Elementary Investigation of the Theory of Numbers.* He commented that the nineteen-digit number $2^{30}(2^{31} - 1)$ is "the greatest [perfect number] that will ever be discovered, for, as they are merely curious without being useful, it is not likely that any person will attempt to find one beyond it." He would be surprised to learn that we now know perfect numbers with over nineteen million digits!

1831

The year mathematician Sophie Germain died in Paris, France. If both p and $2p + 1$ are prime, then p is called a Sophie Germain prime (she had proved that the first case of Fermat's last theorem is true for these primes). [Dennis]

Germain (1776–1831)

James A. Garfield is the only President of the United States to be born on a prime date, i.e., month, day, and year are primes (November 19, 1831). [Blanchette]

1861

The American Civil War (also known as the War Between the States) started in 1861. Did you know there are exactly 18 primes less than or equal to 61? [Greer]

1871

$\frac{1871!}{1781!} - 1$ is prime. [Patterson]

1873

Charles Hermite proved that the number e was transcendental in 1873. Perhaps you have to be a mathematical hermit to achieve such a great feat.

1877

The year British mathematician J.W.L. Glaisher made his pioneering study of maximal prime gaps (see Table 5). He is remembered mostly for work in number theory that anticipated later interest in the detailed properties of modular forms.

1901

Psychiatrist Dr. Frasier Crane's door number on the popular television sitcom *Frasier.* It corresponds to the year Sigmund Freud published *The Psychopathology of Everyday Life.* [Rupinski]

1913

The smallest prime p such that the next prime (1931) is a permutation of the digits of p. Andy Edwards introduced the name "Ormiston pairs" for these after his students at Ormiston College (in Queensland, Australia) manually inspected prime lists and found the first few cases. (An Ormiston k-tuple beginning with p is k consecutive primes each of whose digits are permutations of the digits of p.) [De Geest]

The number of letters in the chemical name for tryptophan synthetase A protein. [Byrne]

"The busy world, which does not hunt poets as collectors hunt for *curios.*" F. Harrison (*Webster's Revised Unabridged Dictionary*, 1913)

1931

The year Gödel's incompleteness theorems were published. Recall that Kurt Gödel (1906–1978) used prime numbers to encode each element of a logical statement. [La Haye]

Nineteen 19's followed by thirty-one 31's is a prime with exactly one hundred digits. [De Geest]

Table 5. Maximal Prime Gaps

gap	prime	gap	prime	gap	prime
0	2	221	122164747	581	1346294310749
1	3	233	189695659	587	1408695493609
3	7	247	191912783	601	1968188556461
5	23	249	387096133	651	2614941710599
7	89	281	436273009	673	7177162611713
13	113	287	1294268491	715	13829048559701
17	523	291	1453168141	765	19581334192423
19	887	319	2300942549	777	42842283925351
21	1129	335	3842610773	803	90874329411493
33	1327	353	4302407359	805	171231342420521
35	9551	381	10726904659	905	218209405436543
43	15683	383	20678048297	915	1189459969825483
51	19609	393	22367084959	923	1686994940955803
71	31397	455	25056082087	1131	1693182318746371
85	155921	463	42652618343	1183	43841547845541059
95	360653	467	127976334671	1197	55350776431903243
111	370261	473	182226896239	1219	80873624627234849
113	492113	485	241160624143	1223	203986478517455989
117	1349533	489	297501075799	1247	218034721194214273
131	1357201	499	303371455241	1271	305405826521087869
147	2010733	513	304599508537	1327	352521223451364323
153	4652353	515	416608695821	1355	401429925999153707
179	17051707	531	461690510011	1369	418032645936712127
209	20831323	533	614487453523	1441	804212830686677669
219	47326693	539	738832927927		

For example, the next prime after 7 is 11; so 7 is followed by a gap formed by 3 composites. This gap is larger than any that preceded it, so it is a maximal prime gap. There are primes that are separated by gaps of 9 and 11, but they follow a gap of 13, so cannot be maximal gaps.

1951

The year Miller and Wheeler used the Electronic Delay Storage Automatic Calculator (EDSAC) to discover a 79-digit prime—the largest known at the time. [Rupinski]

Ralph H. Baer ("the Father of Video Games") came up with the concept of a game machine hooked to a TV as early as 1951.

1973

The smallest prime formed by using all possible end digits of multidigit primes, i.e., 1, 3, 7, and 9. [Gupta]

N.J.A. Sloane's *A Handbook of Integer Sequences* was first published by Academic Press in 1973.

The least prime p that becomes composite if we add or subtract any power of 2 less than $\frac{p}{2}$. (The next few are 3181, 3967, 4889, 8363, 8923, ...)

1987

The Fermat Prize for Mathematics Research started in 1987. It rewards the work of one or more mathematicians in fields where the contributions of Pierre de Fermat have been decisive: statements of variational principles, foundations of probability and analytical geometry, and number theory.

1993

The smallest prime p that gives a zeroless pandigital number when the Fibonacci-like recurrence $a(n) = a(n-1) + a(n-2)$ with $a(1) = 1$ and $a(2) = p$ is applied. [De Geest]

1997

Prime numbers and Cartesian coordinates play a key role in the 1997 movie *Cube*.

The Internet's first general-purpose distributed computing project (`www.distributed.net`) was founded in 1997. Among its first projects was cracking RSA and DES encryption keys.

1999

$1999 = 2^{11} - 7^2$. Note that any positive integer n that obeys the relationship $n^{11} - 7^n$ is called a "convenience store number." Do you see why? [Pomerance]

The least prime number such that the sum of its digits is a perfect number.

The adventures of the TV show *Space: 1999* begin when the Moon is hurled out of Earth's orbit, into deep space.

The first **megaprime** was discovered in 1999. The website *Prime Curios!* was launched the same year. [Doyle]

2003

The smallest "prime year" of the 21st century. The next prime years are 2011, 2017, 2027, 2029, 2039, 2053, 2063, 2069, 2081, 2083, 2087, 2089, and 2099. [Park]

The first edition of *The Music of Primes* by Marcus du Sautoy was published in 2003. [Post]

2011

The next prime date (month, day, and year are primes) from now is 2/2/2011. Note that 222011 is prime too. [Blanchette]

Solar researchers are predicting that the next solar maximum, expected to arrive in 2011, will be the strongest in a half-century.

2017

A total solar eclipse takes place in the continental United States on August 21, 2017. The longest duration of totality will occur over Christian County, Kentucky.

2029

The near-Earth Asteroid "99942 Apophis" caused a brief period of concern in December 2004 because initial observations indicated a 2.7% chance that it would strike the Earth in 2029. Additional observations provided improved predictions that eliminated the possibility of an impact on Earth or the Moon.

2039

A full moon will occur on Halloween in 2039. Primal instincts will be at a maximum on this creepiest night of the century.

2053

Stargazers will get a rare triple planetary treat in November 2053, when Mercury, Mars, and Jupiter all lie within a circle of less than one degree in diameter. Recreational math circles will no less be celebrating the fact that $2053\# - 1$ is a **primorial prime**.

2069

The smallest number that requires a dozen steps to reach a palindrome with the reverse-then-add process.

The next planetary occultation of the star Zavijava (β Virginis) by Venus will take place in August 2069.

2081

The first year of the next "prime decade" (formally called a prime quadruple) that will contain the maximum of four prime years. This will not occur again until the 3250's.

2083

$2083 = (7! - 6! - 5! - 4! - 3! - 2! - 1! - 0!)/2$. [Vatshelle]

2099

The next self prime year is 2099. A self number (or Columbian number) is an integer which cannot be generated by any other integer added to the sum of its digits (e.g., 25 is not a self number because $25 = 17 + 1 + 7$). These were first described in 1949 by the Indian mathematician D. R. Kaprekar.

2111

The square root of 4456321. (The reversal of 4456321 is 1112 squared.)

The smallest "prime year" of the 22nd century. The next prime years are 2113, 2129, 2131, 2137, 2141, 2143, 2153, 2161, and 2179.

August 2006 saw the first outbreak (in U.S. history) of salmonella associated with peanut butter (it was caused by a leaky roof in a

ConAgra Foods plant in Sylvester, Georgia). The only jars affected had a product code beginning with the number "2111" on the lid.

2131

The number of consecutive major league baseball games played by Baltimore Oriole Cal Ripken, Jr., to set a new record. [Litman]

2141

The end of the world! A humorous old puzzle recounts that "Professor Euclide Paracelso Bombast Umbugio of Guayazuela" noted (on April 1, 1946) that the numbers $1492^n - 1770^n - 1863^n + 2141^n$ $(0 \leq n \leq 1945)$ were all divisible by 1946. "Now, the numbers 1492, 1770, and 1863 represent memorable dates: The Discovery of the New World, the Boston Massacre, and the Gettysburg Address. What important date may 2141 be? That of the end of the world, obviously." The puzzle is to "deflate the professor" by explaining why it is easy to find such coincidences. Can you do it? [Trigg]

The smallest prime that differs from its reversal by both a square and cube (hence a sixth power). [Poo Sung]

2143

The process of "slipping" from the Age of Pisces to the Age of Aquarius requires 2143 years. [Skinner]

2153

In 1977, Akio Suzuki used primes ranging from 53 to 2153 to construct the magic cube with smallest possible magic sum (for cubes with side three and distinct prime entries). [Heinz]

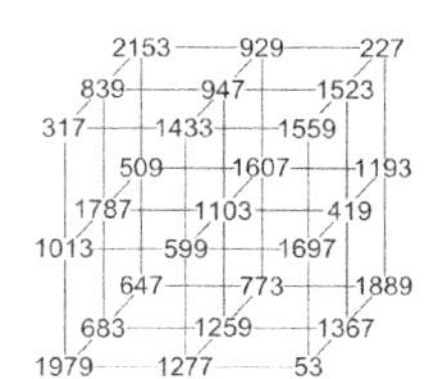

2207

The largest known prime Lucas number to have a composite index. [Dobb]

2213

An "odd" sum of cubes: $2^3 + 2^3 + 13^3$. [Trotter]

2221

The next prime happy-go-lucky year will be 2221. The number 2221 is simultaneously a happy number (page 14) and a lucky number (page 9).

2237

The smallest four-digit prime whose digits are all primes. [Muller]

2251

ThinkGeek, Inc. sells a T-shirt with e on it in which the first 2251 digits of e are used to construct e itself.

2333

Did you miss it in 1801? The planets Mercury through Pluto were arranged east of the Sun in Earth's sky in their "natural order" (from closest to the Sun to the most distant). Don't worry, maybe you can see it next time (in April 2333). [Collins]

2339

There are 2339 ways to arrange the set of twelve pentominoes (5-polyominoes) into a 10-by-6 rectangle, excluding trivial variations obtained by rotation and reflection of the whole rectangle, but including rotation and reflection of a subset of pentominoes. This case was first solved in 1960 by C. B. Haselgrove and Jenifer Haselgrove (now Jenifer Leech).

2357

The smallest prime that contains all of the prime digits.

Chinese Emperor Yao began his reign in 2357 B.C. (A famous legend holds that he created the game of Go to improve the intelligence of his son.)

Twenty-three fifty-seven (23:57) is the largest "prime time" of day on a 24-hour clock in hours and minutes. [Luhn]

$2^2 + 3^3 + 5^5 + 7^7$ is prime. [Papazacharias]

The product of the primes less than or equal to 2357 is the smallest titanic primorial number.

The 1965 Warner Brothers Picture *The Great Race* claims to have the biggest pie fight in cinema history: 2357 pies.

Oklahoma is the only U.S. state name whose letters in prime positions are all consonants.

2441

The first occurrence of the beast number (666) begins at the 2441st digit in the decimal expansion of π. [Gupta]

2477

When the British troop ship HMT *Lancastria* was sunk off Saint-Nazaire, France, there were 2477 survivors.

2503

The number of games played by Babe Ruth in his professional baseball career. [Blanchette]

2521

$2521^2 = 1456^3 - 1455^3$, i.e., one solution to the Diophantine equation $z^2 = x^3 - y^3$. A Diophantine equation is a polynomial equation in which only integer solutions are allowed. These are named after the Greek algebraist Diophantus who lived in the 3rd century. [Mizuki]

2551

The smallest prime p such that $p^2 + p + 4$ is a palindrome. [Russo]

2657

The American mathematician Lowell Schoenfeld (1920–2002) proved that if the Riemann hypothesis was true, then

$$|\pi(x) - \mathrm{li}(x)| \leq \frac{\sqrt{x}\log x}{8\pi}, \text{ for all } x \geq 2657.$$

(See Figure 25 on page 57.)

2711

The memorial to the Holocaust victims of Europe in Berlin consists of 2711 pillars (or stelae) of different heights through which visitors can wander. [Lintermanns]

Waldo gets on school bus number 2711 in the music video "Hot for Teacher" by Van Halen.

2731

If 2731 people are chosen at random, then the probability that at least 17 of them share the same birthday is greater than 50%. By chance, 2731 is a **Wagstaff prime**, i.e., a prime of the form $\frac{2^p+1}{3}$. Note that $\frac{2^{17}+1}{3}$ is also a Wagstaff prime.

2753

There were 2753 episodes from the original NBC Daytime version of *Jeopardy!*.

$|36n^2 - 810n + 2753|$ is prime for $0 \leq n \leq 44$. The values of this polynomial are never divisible by a prime less than 59. [Fung and Ruby]

2777

$2777 = 12 + 23 + 35 + 47 + 511 + 613 + 717 + 819$, where each addend is the concatenation of n and the nth prime. (2777 is the smallest prime formed in this manner.) [Poo Sung]

2851

Chess grandmaster Garry Kasparov's 2851 Elo rating in the July 1999 FIDE rating list is the highest rating ever achieved.

2999

The fourth term of the sequence $2, 11, 29, 2999, \ldots$, where each term (starting with the first prime) is the smallest prime, greater than the previous term, which contains a sum of digits equal to the previous term. The fifth term $(5 \cdot 10^{333} - 10^{332} - 10^{174} - 1)$ was found by Jim Fougeron of Omaha, Nebraska.

Start with any number greater than 1, and write down all its divisors, including 1 and itself. Now take the sum of the individual digits of these divisors. After repeating the process, you'll eventually get the product of the first two odd primes (fifteen) or what Dr. Michael W. Ecker describes as a "mathemagical black hole" with respect to this particular iteration. Note that Jason Earls of Blackwell, Oklahoma, found that 2999 is the least prime that takes exactly fifteen steps to reach the fifteen black hole.

3001

Arthur C. Clarke wrote a book entitled *3001: The Final Odyssey* (published by Del Rey, 1997). [Hartley]

3011

The next year that will be an emirp. [Farrell]

3061

The mean prime gap up to 3061 is the first perfect number: 6 (see Table 11 on page 184).

3119

The number of days between the previous terrorist attack on the Twin Towers and 9/11. Note the appearance of 911 in the reversal of 3119.

3121

The smallest solution to the famous puzzle that first appeared in Ben Ames Williams's "Coconuts" from *The Saturday Evening Post* (October 9, 1926). [Poo Sung]

3137

Delete any digit of 3137 and it will remain prime. [Blanchette]

3187

3187 divides 25496. Each digit is used once, and only once, if a remainder of 0 is included. [Honaker]

3257

The first emirp that contains all the distinct prime digits. [Beedassy]

3301

The smallest of three consecutive prime numbers that are also consecutive lucky numbers. [Harvey]

3313

The smallest prime such that every digit d in it appears exactly d times. [Rivera]

$\pi(3313)$ was the hull number of USS *Prime* (a Vietnam War era minesweeper).

3319

Two raised to the power 3319 is the smallest thousand-digit number of the form 2^n. [Gupta]

3343

The first prime mentioned in *The Five Hysterical Girls Theorem*, Rinne Groff's comeuppance for the fifth grade teacher who threatened to flunk her in math. The cutesy name for the theorem comes from the fact that (when written in the right font) certain prime numbers look like "eeheeheee" when viewed upside down.

3413

A kilowatt-hour of electricity can provide 3413 BTU's of heat. [Bobick]

$3413 = 1^1 + 2^2 + 3^3 + 4^4 + 5^5$. [Patterson]

3511

The largest known Wieferich prime: 3511^2 divides $2^{3511-1} - 1$.

3517

The first three Fermat primes $(3, 5, 17)$ form a prime when concatenated forwards or backwards. They also form primes when concatenated forwards or backwards with a zero inserted between each number. [Vouzaxakis and Hartley]

3527

The city of Karachi, Pakistan, covers 3527 square kilometers.

The number of ways to fold a strip of ten blank stamps. (For example, the five ways of folding a strip of four stamps is in Figure 42.) Note that the digits of 3527 are distinct primes. [Beedassy]

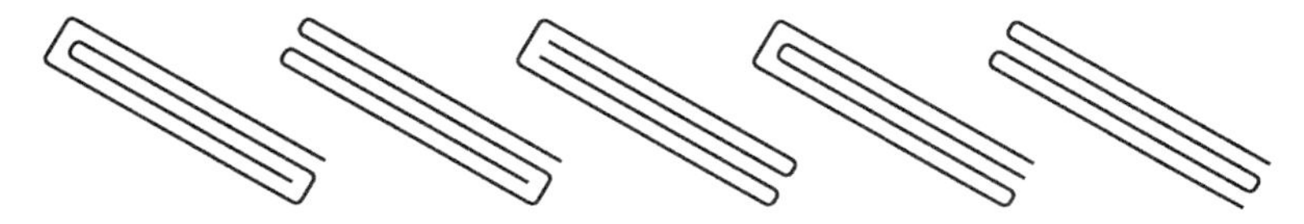

Figure 42. Five Ways to Fold Four Stamps

3539

$C_3H_5N_3O_9$ is nitroglycerin. [Dockery]

3547

The RMS *Titanic* was built to carry 3547 people. [Jackson]

3557

A prime obtained by concatenating the least two pairs of twin primes. [Necula]

3671

The smallest prime factor of the repunit containing 367 1's. [Firoozbakht]

3691

The smallest multidigit prime p such that $\pi(p)$ and the pth prime are palindromic numbers. [Firoozbakht]

3761

The first year of the Hebrew Calendar is 3761 B.C.

3793

The smallest prime whose cube is zeroless pandigital, i.e., contains all digits from 1 to 9. [Gupta]

3797

According to the 16th century prophet Nostradamus, the Earth will survive until A.D. 3797. [Skinner]

11	3851	9257	1747	6481	881	5399
6397	827	5501	71	3779	9221	1831
3881	9281	1759	6361	911	5417	17
839	5381	101	3797	9227	1861	6421
9311	1777	6367	941	5441	29	3761
5387	131	3821	9239	1741	6451	857
1801	6379	821	5471	47	3767	9341

Figure 43. A Prisoner's Prime Magic Square

The center of a 7-by-7 magic square of primes (Figure 43) which stays a magic square (but with 7 primes) if the units digits of all entries are removed (i.e., 9341 becomes 934, . . .). [Madachy]

3803

The reflectable emirp 3803 is the largest prime factor of 123456789. [Bottomley]

4021

In the beginning of "Hot Zone" (an episode of the science fiction television series *Stargate Atlantis*), during Dr. Zelenka's and McKay's game of "Prime or Not Prime," Dr. Zelenka erroneously answers that 4021 is not prime, even though it is.

4027

The only prime permutation of the digits of the smallest composite Mersenne number $2^p - 1$ with p prime, i.e., 2047.

4093

The sum of 4093 and the next consecutive prime can be written as a power of 2. [Rivera]

In 1961, the X-15 rocket plane achieved a world record speed of 4093 miles per hour. [Jenkins]

4159

$4159^2 - 1$ is one of Ramanujan's highly composite numbers. Note that the famous Hardy-Ramanujan number (page 28) is the first four digits of 4159^2. [Gupta]

The arithmetic mean of the two successive Mersenne primes $2^7 - 1$ and $2^{13} - 1$. It is the largest known prime with this property. [Earls]

4201

The smallest prime whose reversal is a tenth power ($1024 = 2^{10}$). [Gupta]

4219

Sound travels through hydrogen at approximately 4219 feet per second (0 °C, 1 atm). [Smith]

4253

$M(4253)$ is the smallest titanic Mersenne prime.

In 1961, Alexander Hurwitz used an IBM 7090 to discover two new Mersenne numbers: $M(4253)$ and $M(4423)$. Because of the way the computer output was stacked, he knew about $M(4423)$ first. So John Selfridge asked, "Does a machine result need to be observed by a human before it can be said to be 'discovered'?" If the answer is yes, then $M(4253)$ was never the largest known prime. [Rupinski]

4259

There are exactly 4259 odd left-truncatable primes. [Angell and Godwin]

4397

Comet Hale-Bopp is predicted to return in the year 4397. The Heaven's Gate cult claimed there was an alien spaceship following behind it.

4493

ThinkGeek, Inc. sells a T-shirt with the π symbol on it in which the first 4493 digits of π are used to construct the π symbol itself.

4567

The only prime number with four consecutive increasing digits. [De Geest]

4649

The smallest composite-digit prime whose prime factors of its digits form another prime when they are concatenated left-to-right. [Honaker]

The greatest prime factor of any seven-digit repdigit. [La Haye]

4663

There are 4663 dodecominoes (12-polyominoes) with holes.

4783

There are 4783 space lattice groups in 4-D Euclidean space. [Post]

4831

Vivian "Sailor Joe" Simmons, a Canadian tattoo artist, died with 4831 tattoos on his body.

4957

Asteroid 4957 Brucemurray is named after Bruce Murray, cofounder of The Planetary Society.

4973

The 666th prime. If you were to intersperse the digits of 666 into this number (i.e., 4696763), then another prime would be formed. [Patterson]

5039

This prime can be written as $1 \cdot 1! + 2 \cdot 2! + 3 \cdot 3! + 4 \cdot 4! + 5 \cdot 5! + 6 \cdot 6!$, or as $7! - 1$. [Sandri]

Table 6. The 2-by-k GAP's of Distinct Primes
(those with the least last terms)

k	GAP	Last Term
2	$3 + 8i + 2j$	13
3	$7 + 24i + 6j$	43
4	$5 + 36i + 6j$	59
5	$11 + 96i + 30j$	227
6	$11 + 42i + 60j$	353
7	$47 + 132i + 210j$	1439
8	$199 + 3300i + 210j$	4969
9	$199 + 3300i + 210j$	5179

Prime for $0 \leq i \leq 1$, $0 \leq j \leq k - 1$

5077

The first term of six prime numbers in arithmetic progression with a common difference of 9876543210. [Honaker]

The smallest conductor of a rank 3 elliptic curve.

5171

The International Correspondence Chess Federation numeric notation for White castling kingside. [McCranie]

5179

A generalized arithmetic progression of primes (a GAP) is a set of integers of the form $a + n_1b_1 + n_2b_2 + \ldots + n_db_d$ with $a, b_1, b_2, \ldots, b_d$ fixed; and the n's run through some range. Table 6 shows the smallest 2-by-k GAP's (those with the least last term) for small k. When k is 9, the largest term is 5179. [Granville]

5273

The sum of the first and only even palindromic prime and the first $5 + 2 + 7 + 3$ odd palindromic primes. Note that the sum is composed of all four palindromic prime digits. [Post]

5381

A term in a decreasing sequence of primes leading to the first prime number, involving the recursive use of the prime counting function: $\pi(5381) = 709, \pi(709) = 127, \ldots, \pi(3) = 2$.

5387

The smallest titanic Fibonacci prime is fib(5387). [Gupta]

5437

The smallest prime whose cube is pandigital (i.e., containing all digits from 0 to 9). [Gupta]

1153	**8923**	**1093**	**9127**	**1327**	**9277**	**1063**	**9133**	**9661**	**1693**	**991**	**8887**	**8353**
9967	8161	3253	2857	6823	2143	4447	8821	8713	8317	3001	3271	**907**
1831	8167	**4093**	**7561**	**3631**	**3457**	**7573**	**3907**	**7411**	**3967**	**7333**	2707	**9043**
9907	7687	**7237**	6367	4597	4723	6577	4513	4831	6451	**3637**	3187	**967**
1723	7753	**2347**	4603	**5527**	**4993**	**5641**	**6073**	**4951**	6271	**8527**	3121	**9151**
9421	2293	**6763**	4663	**4657**	9007	1861	5443	**6217**	6211	**4111**	8581	**1453**
2011	2683	**6871**	6547	**5227**	1873	**5437**	9001	**5647**	4327	**4003**	8191	**8863**
9403	8761	**3877**	4783	**5851**	5431	9013	1867	**5023**	6091	**6997**	2113	**1471**
1531	2137	**7177**	6673	**5923**	**5881**	**5233**	**4801**	**5347**	4201	**3697**	8737	**9343**
9643	2251	**7027**	4423	6277	6151	4297	6361	6043	4507	**3847**	8623	**1231**
1783	2311	**3541**	**3313**	**7243**	**7417**	**3301**	**6967**	**3463**	**6907**	**6781**	8563	**9091**
9787	7603	7621	8017	4051	8731	6427	2053	2161	2557	7873	2713	**1087**
2521	**1951**	**9781**	**1747**	**9547**	**1597**	**9811**	**1741**	**1213**	**9181**	**9883**	**1987**	**9721**

Figure 44. A Prisoner's Prime Magic Square

Figure 44 shows a 13-by-13 bordered magic square made solely of prime numbers. "Bordered" because if the outer numbers are removed, the result is a magic 11-by-11 square. And nested within that, 9-by-9, 7-by-7, 5-by-5, and 3-by-3 magic squares all centered on 5437. Who would have time to create such a marvelous square? In this case, a man in jail. [Madachy]

5557

$2 + 3 + 5 + 7 + 11 + \ldots + 3833 = 3847 + 3851 + \ldots + 5557$. [Rivera]

5711

The smallest prime that is a concatenation of three consecutive primes (5, 7, and 11). [De Geest]

5737

The start of the first set of six consecutive full period primes.

5741

The fourth prime Pell number. The Pell numbers P_n begin 0, 1; and then satisfy the recurrence relation $P_{n+1} = 2P_n + P_{n-1}$. For a Pell number to be prime, it is necessary that n be prime.

5813

The smallest prime formed with three consecutive Fibonacci numbers in ascending order. [Gallardo]

5851

The only known prime whose sum of digits is shared with that of its square and cube. [Beedassy]

5857

In January 1961, "Ham the Astrochimp" blasted off from Kennedy Space Center in Florida and attained a velocity of 5857 miles per hour on a Mercury spacecraft before splashing down safely in the Atlantic Ocean, thus achieving the Mission Objective: primate suborbital and auto abort.

5879
The start of a record-breaking run of eleven consecutive integers with an odd number of prime factors. [Gupta]

5881
The largest member in the first pair of twin primes separated from the next pair by over two hundred.

6079
The sum of the odd primes in Benjamin Franklin's original 16-by-16 square (see Figure 45).

Franklin (1706–1790)

6089
The start of a sequence of circular-digit primes. The terms consist of primes of the form $609 \cdot 10^n - 1$, for n from 1 to 6. [Earls]

6101
The largest non-**titanic prime** is $10^{999} - 6101$. Note that 6101 is an invertible prime. [Boivin]

6173
Delete any digit of 6173 and it will remain prime. [Blanchette]

6421
The sum of the first 6421 non-composites is a perfect square.

6427
The smallest prime formed from the reverse concatenation of two consecutive cubes. [Gupta]

6481
The smallest prime formed from the concatenation of two consecutive squares. [Gupta]

6521
$\pi(6521) = 6! + 5! + 2! + 1!$. There is only one other prime number with this property. [Honaker]

200	217	232	249	8	25	40	57	72	89	104	121	136	153	168	185
58	39	26	7	250	231	218	199	186	167	154	135	122	103	90	71
198	219	230	251	6	27	38	59	70	91	102	123	134	155	166	187
60	37	28	5	252	229	220	197	188	165	156	133	124	101	92	69
201	216	233	248	9	24	41	56	73	88	105	120	137	152	169	184
55	42	23	10	247	234	215	202	183	170	151	138	119	106	87	74
203	214	235	246	11	22	43	54	75	86	107	118	139	150	171	182
53	44	21	12	245	236	213	204	181	172	149	140	117	108	85	76
205	212	237	244	13	20	45	52	77	84	109	116	141	148	173	180
51	46	19	14	243	238	211	206	179	174	147	142	115	110	83	78
207	210	239	242	15	18	47	50	79	82	111	114	143	146	175	178
49	48	17	16	241	240	209	208	177	176	145	144	113	112	81	80
196	221	228	253	4	29	36	61	68	93	100	125	132	157	164	189
62	35	30	3	254	227	222	195	190	163	158	131	126	99	94	67
194	223	226	255	2	31	34	63	66	95	98	127	130	159	162	191
64	33	32	1	256	225	224	193	192	161	160	129	128	97	96	65

Figure 45. A 16-by-16 Franklin Square

Benjamin Franklin (1706–1790) stated that it is "the most magically magical of any magic square ever made by any magician." Each row and column add to 2056, as do the entries in any 4-by-4 subsquare. The "bent diagonals" also add to 2056, and the half-rows and half-columns add to 1028. Despite this, it is not a true magic square as the main diagonals do not add to the sum 2056. (This square was first mentioned in a letter circa 1752 and first published in 1767.)

6563

The largest known prime of the form $3^{2^n} + 2$.

6569

The smallest prime p such that p^{11} is equal to the sum of 11 consecutive primes. [Rivera]

6661

The smallest **beastly prime** is also an invertible prime (1999).

6689

The smallest invertible prime whose digits are circular and composite. [Punches]

The NSW number with index 6689 is prime. These numbers arise when addressing the following question: "Is there a finite simple group whose order is a square?"

6709

The smallest prime factor of the 43rd Lucas number. Note that the smallest prime factor of the next composite Lucas number with prime index can be obtained by deleting the first digit of 6709.

The largest in the set of primes that appear in at least one of the nine basic solutions (not counting rotations of 120 degrees or reflections) to the Tetractys Puzzle (see page 100): 2, 3, 5, 7, 13, 17, 23, 29, 31, 37, 41, 43, 47, 53, 59, 61, 67, 71, 73, 79, 83, 89, 97, 179, 239, 241, 251, 269, 271, 281, 283, 347, 359, 389, 479, 541, 569, 571, 593, 673, 709, 743, 821, 823, 853, 907, 953, 971, 983, 2389, and 6709.

6793

The start of the first occurrence of three consecutive primes ending with the digit three (6793, 6803, 6823). [Murthy]

6823

The Tetragrammaton (YHWH) is one of the names of the God of Israel and occurs 6823 times in the Old Testament (according to the Jewish Encyclopedia). [Earls]

6899

The smallest prime whose sum of digits is a fifth power. [Gupta]

6991

In the movie *Die Hard: With a Vengeance*, John McClane's colleague Ricky Walsh always bets his badge number (6991) as a lottery number. [Poo Sung]

7039

The world's largest cocktail was a Margarita measuring 7039 gallons. It was made in 2001 by staff at Jimmy Buffett's Margaritaville and Mott's Inc., Universal City Walk in Orlando, Florida.

7057

In Mark Haddon's 2003 debut novel, *The Curious Incident of the Dog in the Night-time*, Christopher John Francis Boone is an autistic teenager who has memorized all the world's countries and their capital cities as well as every prime number up to 7057. [Wichmann]

7103

This prime can be formed using the first five primes: $\frac{2^3+5^7}{11}$. [Brown]

7109

The longest recorded prison sentences were given in 1969 to two "confidence tricksters" in Iran: 7109 years (from *Norris and Ross McWhirter's 10 Best Oddities and Fun Trivia*).

7121

The sum of the first twenty palindromic primes. [Post]

7129

A prime that can be written as $9^4 + 8^3 + 7^2 + 6^1 + 5^0$. Each digit appears once, and only once. [Wagler]

One of the mysterious numbers found in Ed Leedskalnin's bedroom at Coral Castle Tower (near Homestead, Florida). Did this eccentric man rediscover the secrets of ancient pyramid building as he claimed? [Schuler]

7207

National Football League quarterback Peyton Manning finished his high school career with a number of passing yards that is both an emirp and a lucky number: 7207.

7213

The sum of the odd primes up to 7213 is a palindromic prime. [De Geest]

7297

The diagonals of a regular 24-gon intersect at 7297 internal points (see Figure 46).

7321

The smallest zeroless prime whose product of the first two digits is a concatenation of the 3rd and 4th digits. [Gallardo]

7331

The smallest prime whose conversion from seconds into hours, minutes, and seconds, gives a prime number of hours, minutes, and seconds ($7331s = 2h\ 2m\ 11s$). [Thoms]

7351

The product of the factorials of the digits of 7351 produce another factorial: $7! \cdot 3! \cdot 5! \cdot 1! = 10!$.

The sum of the digits of 7351 equals the square of their average. Note that the first four odd numbers are used. [Edwards]

7523

The largest left-truncatable prime that contains only distinct prime digits. [Patterson]

7669

The number of twin lucky numbers less than a million. [Schneider]

David MacQuarrie's "PrimeTimeClock" will only display the time (as *hours:minutes:seconds*) if the corresponding number is prime. It displays 7669 different times each day and indicates whether a prime

Figure 46. A Prime Number of Intersections

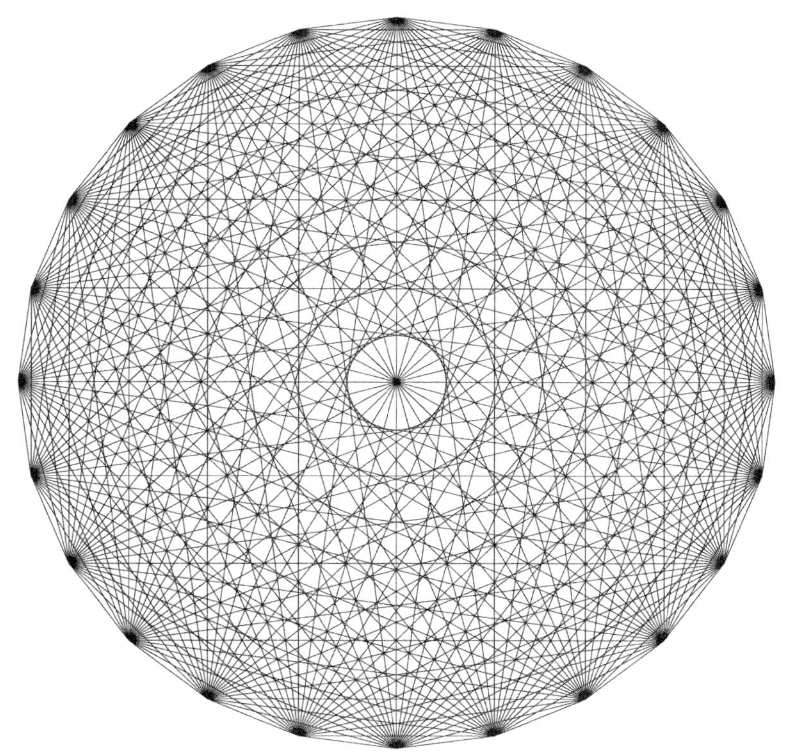

The diagonals of a regular 24-gon (above) intersect at 7297 internal points (6144 intersections of two diagonals, 864 of three, 264 of four, 24 of five, and one intersection of twelve diagonals).

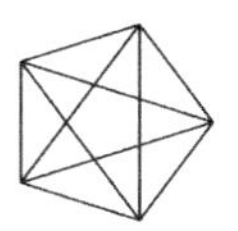

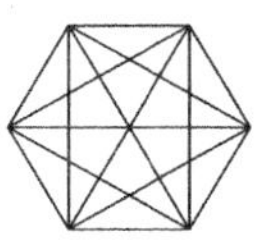

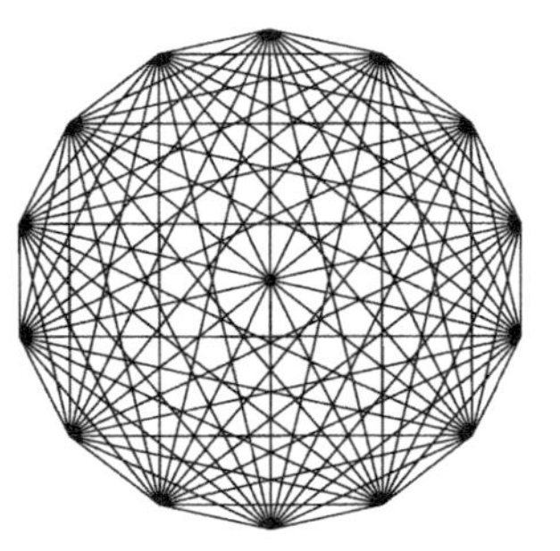

There are three smaller examples where the number of intersections is also prime: the pentagon, hexagon and the 14-gon. Some larger examples include the regular polygons with 44, 58, 72, 76, 80, 84, 86, 104, 128, 134, 138, 180, 186, 188, 218, 228, 246, 256, 266, 280, 300, 320, 352, and with 360 sides. This last one has 677630881 internal intersections.

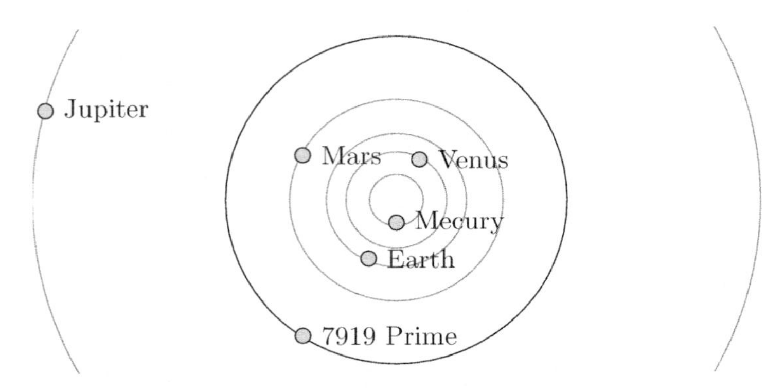

Figure 47. The Orbit of Asteroid "7919 Prime" (1981 EZ27)

is special (e.g., a Sophie Germain prime) by illuminating its beacon in different colors.

7877
An absolute prime in base nine.

7919
The name of Asteroid 7919 is *Prime*, literally (Figure 47). 7919 is the one thousandth prime number. [Gallardo]

7927
The Earth would just fit through a 7927-mile diameter hoop. [Punches]

7951
An iccanobiF emirp.

7963
The largest zeroless prime whose product of the first two digits is a concatenation of the 3rd and 4th digits. [Capelle]

7993
The sum of 6023 and 1970. It was used by Charles Babbage in

his autobiography *Passages From the Life of a Philosopher* to explain the process of addition. [Haga]

Babbage (1791–1871)

8087

The "8087" was the first math coprocessor designed by Intel. Its purpose was to speed up computations involving floating-point arithmetic. [Gallardo]

8101

The $(1018 + 1)$th prime. [Bopardikar]

8161

$8161 = 2^{13} - 31$. (13 and 31 are emirps.)

8191

This number turned upside down forms the first four digits in the decimal expansion of the golden ratio: $\phi = \frac{1+\sqrt{5}}{2} = 1.618....$ [Wu]

There is only one prime less than 8191 that is also a repunit in three bases. Can you find it? [Pimentel]

The smallest Mersenne prime p such that the Mersenne number $M(p) = 2^p - 1$ is composite.

8363

The number of five-digit prime numbers. [Dobb]

8513

President Theodore Roosevelt still holds the Head of State handshaking record—he shook 8513 hands on New Year's Day, 1907.

8609

The largest circular-digit prime whose digits are distinct. Note that 6089 and 8069 are also primes. [Trotter]

8731

Sherlock Holmes' address 221B is the hexadecimal representation of the prime 8731. [Rupinski]

8741
In Oliver Sacks' book *The Man Who Mistook His Wife for a Hat*, C. C. Park asks a young autistic man named "Joe" if there was anything special about 8741. He replied, "It's a prime number."

9007
The RSA public encryption algorithm depends on the how much easier it currently is to find primes, than to factor. To illustrate the RSA's security, an encrypted message was published in the August 1977 issue of *Scientific American*. The message used this prime as an encryption key. To decrypt the message one had to factor a 129-digit number—a task Rivest (the R of RSA) predicted would take 40 quadrillion years. It took a little less than seventeen years! The decrypted message? "The Magic Words are Squeamish Ossifrage." This began a tradition of using the words "squeamish ossifrage" in cryptanalytic challenges. Ossifrage is an older name for the lammergeier (a scavenging vulture).

9013
The largest currently assigned Australian Postcode that is prime. It belongs to the Commonwealth Bank in Brisbane QLD. [Hartley]

9041
The largest prime whose digits are distinct square digits. [Gupta]

9091
A unique prime indeed. Reverse the order of its middle digits to form the next unique prime (9901).

9341
Goldbach's conjecture is usually satisfied with one of the two primes being relatively small. For numbers under 10^{18}, the largest this smaller prime needs to be is 9341. This is part of the reason that there are so many ways to write even integers as the sum of two primes (see Goldbach's Comet on page 149).

9371
The largest prime formed with the four digits that can end multidigit primes. [Capelle]

Figure 48. Goldbach's Comet: the number of ways to write even integers as the sum of two primes

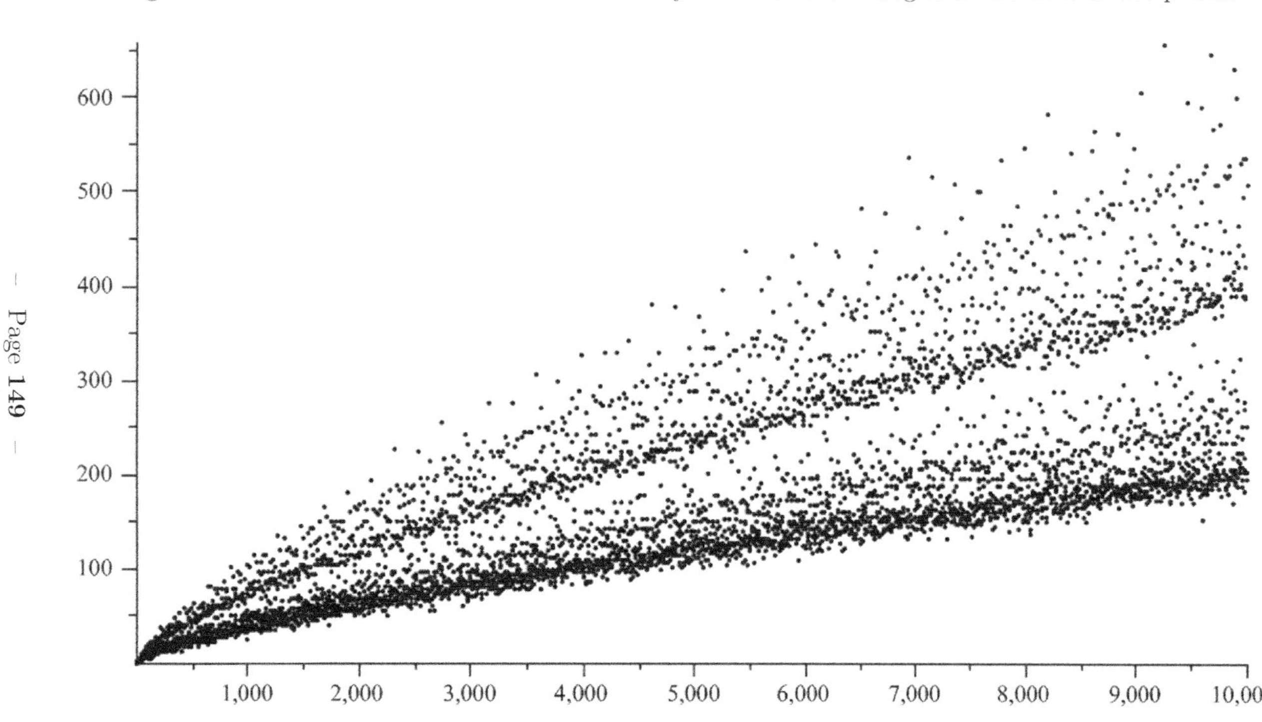

9403

The prime number 9403 divides 65821, 7 times. Each digit appears once, and only once. [Honaker]

9419

The start of four consecutive twin prime pairs. [Rathbun]

9431

$9^2 + 4^2 + 3^2 + 1^2$ is prime. [Trigg]

9551

Compare 9551 with the 9551st prime. [Punches]

9689

The only known multidigit Mersenne prime exponent that starts and ends with the same digit. [Luhn]

9767

The largest 4-digit prime that can be formed by the concatenation of two 2-digit primes (97 and 67). [Sladcik]

9901

Turn this prime number upside down to get the year of the Norman Conquest of England.

A prime whose square (98029801) is the concatenation of two consecutive numbers in descending order. [Bopardikar]

9923

Phil Carmody found that $38 \cdot 256 + 195$ is the smallest executable prime on an `x86 DOS`. (It is the numerical equivalent of the program `ES:RET`, i.e., segment override, which should execute on any `x86` system.)

9949

The largest four-digit prime number that is turned into a composite number if any digit is deleted.

10007

The smallest assigned 5-digit ZIP Code (with no leading zeros) that is prime. It belongs to the Financial District of New York City (sometimes called FiDi). [Baker]

10009

Alphabetically the first prime in Slovenian (desettisocdevet). [Demailly]

10061

The smallest non-assigned 5-digit ZIP Code (with no leading zeros) that is prime. [Punches]

10069

The smallest five-digit prime whose permutations of digits can yield nine distinct perfect squares. [McCranie]

10093

The Little Book of Big Primes (Springer-Verlag, 1991) by Paulo Ribenboim contains a short section called "Primes up to 10,000"—however, the list actually goes up to 10093.

10177

The smallest five-digit prime that produces five other primes by changing its first digit to a nonzero digit. [Opao]

10243

The smallest prime with five distinct digits. [Backhouse]

10427

There are 0 primes less than or equal to 1, 4 primes less than or equal to 10, and 27 primes less than or equal to 104. [Honaker]

10501

A palindromic prime that is the sum of three consecutive primes, while at the same time serving as the middle prime of a set of three consecutive primes whose sum is another palindromic prime. [Trotter]

A pair of twin primes surround $60n^2 + 30n - 30$ for each value of n from 1 to 13. Note that 10501 is the largest prime in this sequence. [Blanchette]

10631

The reverse concatenation of the first four triangular numbers. [Gupta]

10909

On day 10909 of *The Truman Show* starring Jim Carrey, odd things begin happening that slowly convince Truman that the world around him is not as it seems.

10987

The prime sum of the alphametic ALI + BABA + WAS + A = LIBRA. [Faires]

The smallest prime formed from the reverse concatenation of four consecutive numbers. [Gupta]

11213

The Mathematics Department at the University of Illinois had their postage meter changed to stamp "$2^{11213} - 1$ IS PRIME" after a record-breaking Mersenne prime was found there in 1963.

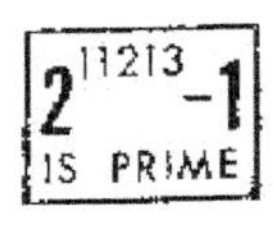

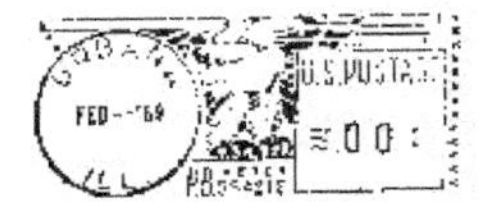

11311

The only five-digit number such that its nth digit equals the remainder when the number is divided by $n + 1$. [Rupinski]

11353

A prime such that the sum of the squares of its digits is equal to the product of its digits. [Russo]

11551

The smallest five-digit prime with limerick rhyme scheme (*aabba*). [Baldwin]

11593

The first in a sequence of nine consecutive primes of the form $4n + 1$. [Haan]

11863

The sum of the odd primes up to 11863 is a palindromic prime. [De Geest]

11939
The only cyclic emirp with five digits.

12601
The largest prime which is not the sum of squares of distinct primes. [Crespi de Valldaura]

12841
The smallest prime such that the sum of its digits cubed is equal to the product of its digits squared. [Earls]

13147
The smallest prime p such that p^2 contains exactly nine different digits.

13297

Table 7. First Prime to Take n Steps

prime	steps	prime	steps
19	2	739	17
59	3	829	10
79	6	1297	21
89	24	2069	12
167	11	10853	26
193	8	13297	29

The smallest prime number that requires 29 steps to reach a palindrome with the reverse-then-add process (Table 7). For some odd reason, the primes often seem to be the first to require the largest number of steps. (See also the example in Table 2 on page 60.)

13327
The sum of Euler's list of 65 "numeri idonei" (listed in Table 8).

13331
Inserting any digit d between adjacent digits of this palindromic prime never produces a new prime. [De Geest]

13597
The smallest prime (emirp) containing all of the odd digits. [Poo Sung and Loungrides]

13679

The first primeval number to contain over one hundred (actually 106) embedded primes. [Keith]

13831

The smallest multidigit palindromic prime that becomes palindromic if added to the next prime. [Trotter]

13831 plus a googol (10^{100}) is prime.

14159

The first prime formed from the decimal expansion of '$\pi - 3$'.

14401

On the syndicated game show *Jeopardy!*, the record-setting winning streak of Ken Jennings ended when opponent Nancy Zerg defeated him with a final dollar amount of $14401. [Rupinski]

14593

The largest prime factor of 12345678. [Cherry]

14741

The smallest palindromic prime that contains all of the straight digits. [Gupta]

14869

The smallest prime that contains all of the composite digits. [Murthy]

Table 8. Euler's "numeri idonei" (see 13327)

1	2	3	4	5	6	7	8	9	10
12	13	15	16	18	21	22	24	25	28
30	33	37	40	42	45	48	57	58	60
70	72	78	85	88	93	102	105	112	120
130	133	165	168	177	190	210	232	240	253
273	280	312	330	345	357	385	408	462	520
760	840	1320	1365	1848					

15359

Let p_i be the ith prime. Then $p_k\# < \binom{k^2}{k}$ for $2 < k < 1794$. But $p_k\# > \binom{k^2}{k}$ for $k \geq 1794$. Just what is this key prime p_k? 15359.

15803

The sum of the cubes of the first 2^3 primes.

16033

The smallest prime to be both preceded and followed by more than a score of consecutive composite numbers.

16361

The smallest prime in Russo's Truncated Palindromic Prime Pyramid.

16361
1163611
311636113
33116361133
3331163611333
333311636113333
Russo's Pyramid

16661

The smallest palindromic beastly prime.

16661 is the 1928th prime. Note that $1 + 6 + 6 + 6 + 1$ equals $1 + 9 + 2 + 8$.

The smallest invertible palindromic prime. [Capelle]

16843

The smallest **Wolstenholme prime**. Wolstenholme's theorem can be stated in many ways, including that if $p \geq 5$, then p^3 divides $\binom{2p}{p} - 2$. Wolstenholme primes are those for which that expression is also divisible by p^4. Here $\binom{2p}{p}$ is the central binomial coefficient.

17203

$(8! - 7! - 6! - 5! - 4! - 3! - 2! - 1! - 0!)/2$. [Vatshelle]

17291

The smallest prime containing the famous Hardy-Ramanujan number 1729 (page 28) as a substring. [Gupta]

17839

On September 3, 2003, Brian Silverman did an Internet search on 5-digit numbers up to 30000. On that day, according to *Google*, 17839 was the least popular of all these numbers.

18041

The smallest prime that starts a prime quadruple, whose digits can be rearranged to form a different prime (i.e., 81041) that starts another prime quadruple. [Puckett]

18731

The smallest prime that is the mean of the two preceding and two following primes. [Dale]

18757

$|3n^3 - 183n^2 + 3318n - 18757|$ is prime for $0 \leq n \leq 46$. The values of this polynomial are never divisible by a prime less than 37. [Martín-Ruiz]

18839

The start of the first occurrence of four consecutive primes ending with the digit nine (18839, 18859, 18869, 18899). [Murthy]

19249

In May 2007, the **Proth prime** $19249 \cdot 2^{13018586} + 1$ became the largest known real Eisenstein prime. Figure 49 shows the distribution of the small Eisenstein primes. (This prime was found as part of the Seventeen or Bust project.)

19541

The smaller member in a pair of five-digit twin primes that can be expressed as the sum of squares of five primes. [Patterson]

19681

An emirp that can be expressed with the first two prime numbers: $(3^3)^3 - 2$. Note that 19681 becomes a semiprime if turned upside down. [Patterson]

19937

The largest known Mersenne prime exponent whose digits are all odd. Bryant Tuckerman found $M(19937)$ using an IBM 360/91 computer in 1971. The number 19937 is also a **circular prime**.

20161

Every number greater than 20161 can be expressed as a sum of two abundant numbers. [Rupinski]

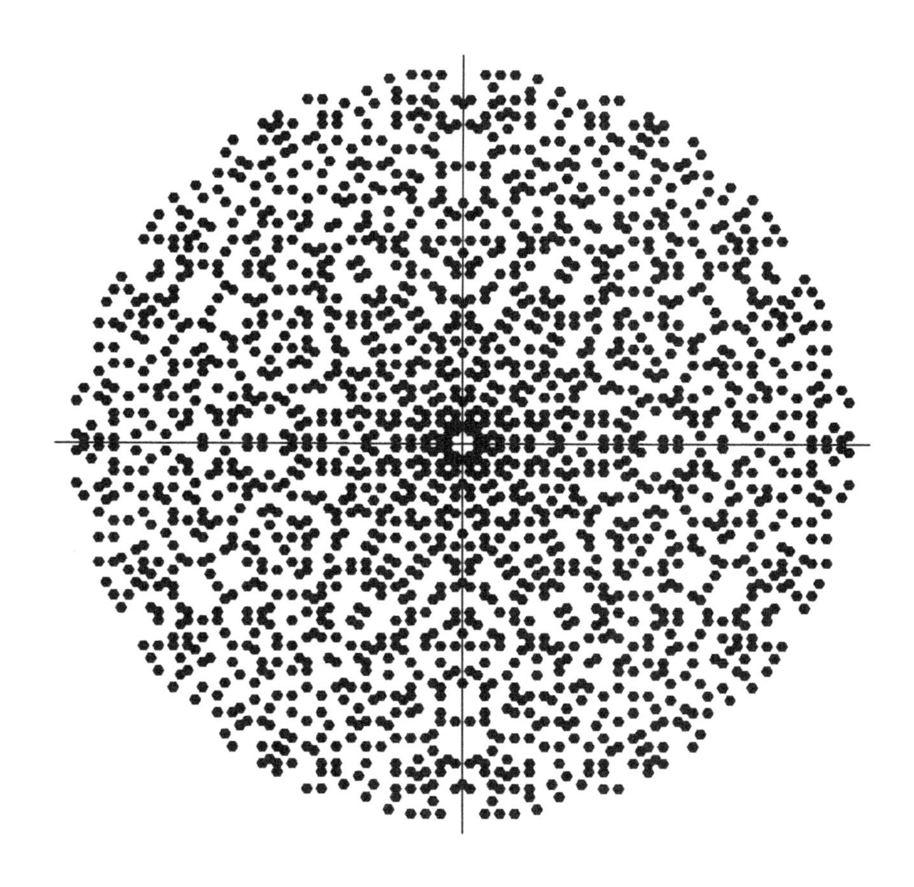

Figure 49. The Small Eisenstein Primes

Eisenstein primes are the primes in the set of numbers of the form $a + bw$, where a and b are integers and $w = \frac{-1+i\sqrt{3}}{2}$, such that $a + bw$ cannot be written as a product of other Eisenstein integers. The center of the image above is 0 and the primes $a + bw$ are indicated by small hexagons placed at $a + bw$ on the complex plane. Eisenstein primes come in three flavors: the prime $1 - w$; the primes $\pm a$, $\pm aw$ and $\pm aw^2$, where a is a real prime congruent to 2 modulo 3; and $a \pm bw$, where $a^2 - ab + b^2$ is a real prime.

20543
The ZIP Code of the U.S. Supreme Court (1 First Street, N.E., Washington, D.C. 20543).

21397
The largest prime whose square contains no duplicate digits: $21397^2 = 457831609$. [Beedassy]

21701
The largest assigned 5-digit ZIP Code that is a Mersenne prime exponent. It belongs to Frederick, MD.

$M(21701)$ made international news when found by high school students Curt Landon Noll (now Landon Curt Noll) and Laura Nickel (now Ariel T. Glenn) using the California State University Cyber 174 on Halloween Eve in 1978. Note that Noll had previously developed a wide collection of "ASCII bats" (e.g., /\../\, /\oo/\, etc.) which have become part of his personal trademark.

22193
The lesser member in the smallest set of three consecutive primes whose sum of digits is prime and equal. [Gupta]

22229
The smallest assigned 5-digit ZIP Code with no leading zeros that is a **near-repdigit prime**. It belongs to Arlington, VA.

22273
The largest prime number in the Bible. It appears in Numbers 3:43. [Blanchette]

22501
The start of the first occurrence of four consecutive primes ending with the digit one (22501, 22511, 22531, 22541). [Murthy]

22963
The start of the first occurrence of four consecutive primes ending with the digit three (22963, 22973, 22993, 23003). [Murthy]

23143
The start of a dozen primes in arithmetic progression with a common difference of 30030. [Golubiev]

23159

Pierre de Fermat lived 23159 days. [Blanchette]

23209

In 1979, Landon Curt Noll discovered his second Mersenne prime $M(23209)$. He was just out of high school at the time.

Noll's License Plate

23333

A right-truncatable prime using the first and second prime. [Capelle]

23819

The deer hunting license number on Robert De Niro's back as he portrays the character Michael Vronsky in *The Deer Hunter* (1978). [Haga]

24133

The sum of the first one hundred prime numbers.

26861

It is at $p = 26861$ that for the first time the primes of the form $4n + 1$ (up to p) outnumber those of the form $4n + 3$. (Their lead is brief, $4n + 3$ primes catch up at 26863, and $4n + 1$ primes do not regain the lead in this prime race until 616841.) [Leech]

27541

According to the FBI's Uniform Crime Reporting Program, the city of Houston, Texas, had 27541 burglaries in 2005. Now that's a *crime number*!

27611

In 1990, Balog proved there are arbitrarily large sets of primes for which the means of each pair of entries is a distinct prime. For example,

$$\{71, 1163, 1283, 2663, 4523, 5651, 9311, 13883, 13931, 14423, 25943, 27611\}.$$

(See Table 9 for more examples.) Granville believes there may be sets like this of infinite length!

Table 9. Prime Sets with Prime Pairwise Means (with Smallest Possible Largest Terms)

n	Set of n Primes
2	$\{3, 7\}$
3	$\{3, 7, 19\}$
4	$\{3, 11, 23, 71\}$
5	$\{3, 11, 23, 71, 191\}$
6	$\{3, 11, 23, 71, 191, 443\}$
7	$\{5, 17, 41, 101, 257, 521, 881\}$
8	$\{257, 269, 509, 857, 1697, 2309, 2477, 2609\}$
9	$\{257, 269, 509, 857, 1697, 2309, 2477, 2609, 5417\}$
10	$\{11, 83, 251, 263, 1511, 2351, 2963, 7583, 8663, 10691\}$
11	$\{757, 1009, 1117, 2437, 2749, 4597, 6529, 10357, 11149, 15349, 21757\}$
12	$\{71, 1163, 1283, 2663, 4523, 5651, 9311, 13883, 13931, 14423, 25943, 27611\}$

27941

The only known prime p such that $n^2 - n + p$ produces exactly 600 primes for $n = 0$ to 1000. [Rodriguez]

30103

The only known multidigit palindromic prime found by averaging the divisors of a composite number. [McCranie and Honaker]

The decimal expansion of the common logarithm of two rounded to five digits. [La Haye]

30689

The smallest prime that contains each of the curved digits (0, 3, 6, 8, and 9). [Gupta]

30757

The smallest assigned 5-digit ZIP Code with no leading zeros that is a Fibonacci prime index. It belongs to Wildwood, GA. This is in Dade County, at the extreme northwestern part of the state, which curiously appears to be left out on the Georgia State Quarter.

31259
The number of people employed at the Consolidated Insurance Company in the Oscar-winning movie *The Apartment* (The Mirisch Corporation, 1960). [Litman]

31337
Hacker lingo for "elite." [Jupe]

31607
The largest prime less than the square root of 10^9. It is the largest prime that is sufficient when using trial division or the Sieve of Eratosthenes algorithm (see Figure 50) for numbers less than 10^9.

32057
Add 2 to any digit of 32057 and it will remain prime. [Blanchette]

32423
The sum of the first 32423 prime numbers is pandigital. No other palindromic number shares this property. [Honaker]

34273
The smallest prime whose square is zeroless pandigital. [Gupta]

34421
The smallest prime expressible in six distinct ways as the sum of consecutive primes. [Rivera]

34543
$\pi(34543) = 3^3 + 4^4 + 5^5 + 4^4 + 3^3$.

34847
The sum of 34847 plus $34847 + 2$ is the only known palindromic square which is the sum of a twin prime pair. [Honaker]

35759
It is impossible to get from the start (room 2) to room number 35759 in Paulsen's Prime Number Maze. Paulsen called 35759 "the first interesting case." The maze is based on the binary representation of the primes: you may change only one binary digit at a time, or add a 1 to the beginning; but may only move from prime to prime. So from the start, room 2 (10_2), you can go to 3 (11_2), then to 7 (111_2) and to 5 (101_2). [Hartley]

Figure 50. Finding the Small Primes with the Sieve of Eratosthenes

The Sieve of Eratosthenes (from about 240 B.C.) is one of the fastest ways to find all of the primes less than a small number. It requires only four cycles through its simple process to find all of the primes less than 49.

43	44	45	46	47	48
37	38	39	40	41	42
31	32	33	34	35	36
25	26	27	28	29	30
19	20	21	22	23	24
13	14	15	16	17	18
7	8	9	10	11	12
	2	3	4	5	6

First, start with an list of integers. Two is the first prime, so circle it and cross out all other multiples of two. As $3 < 2 \cdot 2$, 3 is also prime.

43	44	45	46	47	48
37	38	39	40	41	42
31	32	33	34	35	36
25	26	27	28	29	30
19	20	21	22	23	24
13	14	15	16	17	18
7	8	9	10	11	12
	2	3	4	5	6

Next, the first remaining prime is three, so circle it, and cross out all other multiples of three. Since 5 and 7 are less than $3 \cdot 3$, they are also prime.

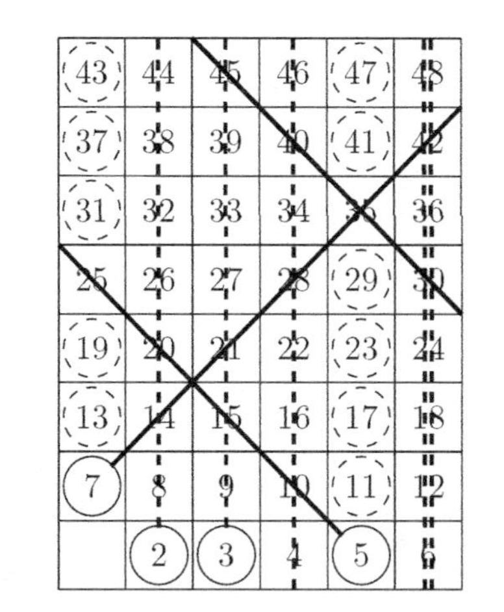

Finally, repeat the process with the first remaining prime, five; and again with the first it leaves, seven. Anything else less than $7 \cdot 7$ is prime.

35803

A satellite is "geostationary" (the period of its circular orbit equals that of the Earth's rotation) at an altitude of approximately 35803 km. The idea was first popularized in a 1945 paper by Arthur C. Clarke.

35899

Each of these differences of factorials in Figure 51 is prime, but sadly the next term is composite. (These terms are also prime if we start at 10, 15, 19, 41, 59, 61, 105, and 160.)

$$\begin{gathered}3!-2!+1!=5,\\4!-3!+2!-1!=19,\\5!-4!+3!-2!+1!=101,\\6!-5!+4!-3!+2!-1!=619,\\7!-6!+5!-4!+3!-2!+1!=4421,\\8!-7!+6!-5!+4!-3!+2!-1!=35899.\end{gathered}$$

Figure 51. Six Prime Differences of Factorials

36263

The smallest prime that can be represented as the sum of a prime and its reversal in four different ways. [Gupta]

39883

The smallest prime that can be represented as the sum of a prime and its reversal in five different ways. [Gupta]

40487

The smallest odd "non-generous" prime p with the property that the smallest primitive root (page 76) of p is not a primitive root of p^2. (Only three non-generous primes are currently known; the others are 2 and 6692367337.) [Glasby]

40609

The smallest prime formed from the concatenation of the first n semiprimes separated by zeros. [Post]

42349

The number held by a hitchhiker on the front cover of Richard L. Francis' book *The Prime Highway.*

43201

The largest five-digit prime formed using every digit smaller than five. [Punches]

45361

$45361 = 8! + 7! + 1!$ and $871 = 4! + 5! + 3! + 6! + 1!$. The only pair of primes with this property. [Hartley]

45823

The start of a sequence of five consecutive primes with the same pattern of consecutive gaps between primes as the sequence of odd primes. [McCranie]

47501

$\pi(47501) = -4! + 7! - 5! + 0! - 1!$.

50069

$1^1 + 2^2 + 3^3 + 4^4 + 5^5 + 6^6$. [Patterson]

51413

A prime number obtained by reversing the first five digits of the decimal expansion of π.

54163

The sum of all prime years in the 20th and 21st century. [Bopardikar]

54973

The smallest prime ending with 3 whose sum of digits equals the sum of digits of its square. [McCranie]

59281

The sum of the first three semiprimes to the powers of the first three primes: $59281 = 4^2 + 6^3 + 9^5$. [Post]

60101

The smallest prime whose reciprocal has period length one hundred.

60659

In the Sci-fi movie *Cube 2: Hypercube* (2002), the key to escaping the cube centers on the recurrence of the prime number 60659. [Rupinski]

61843

The only multidigit prime factor of 8549176320 (a pandigital integer formed from sorting the digits alphabetically in English). [Grantham]

63241

One light-year is about 63241 astronomical units.

63527

Clyde Champion Barrow (of Bonnie and Clyde fame) became prisoner number 63527 at Eastham Prison Farm Number 2 in Texas on April 21, 1930. While incarcerated he had a fellow convict chop off two of his toes with an axe to avoid a work detail. [Post]

63647

The world's longest palindromic sentence created by Peter Norvig in 2002 contained 63647 letters. Not to mention 15139 words.

64037

The Baxter-Hickerson function,

$$(2 \cdot 10^{5n} - 10^{4n} + 2 \cdot 10^{3n} + 10^{2n} + 10^{n} + 1)/3,$$

produces integers whose cubes lack the digit zero. 64037 is the smallest odd prime of that form. [Honaker]

64553

The first of three consecutive primes, each containing their own embedded ordinal number (e.g., 64553 is the 6455th prime). Try your luck at guessing the other two primes. [Gallardo]

65521

The larger member in a sequence of primes formed by taking the product of the first n nonzero Fibonacci numbers and then adding 1.

The largest prime less than 2^{16}. [Noe]

65537

The largest known Fermat prime: $2^{2^4} + 1$. The others are $2^{2^0} + 1$, $2^{2^1} + 1$, $2^{2^2} + 1$, and $2^{2^3} + 1$. It is suspected there may be no more.

To remember the digits of 65537, recite the following mnemonic: "Fermat prime, maybe the largest." Then count the number of letters in each word. [Brent]

The smallest prime that is the sum of a nonzero square and a nonzero cube in four different ways: $122^2 + 37^3$, $219^2 + 26^3$, $255^2 + 8^3$, and $256^2 + 1^3$. [Post]

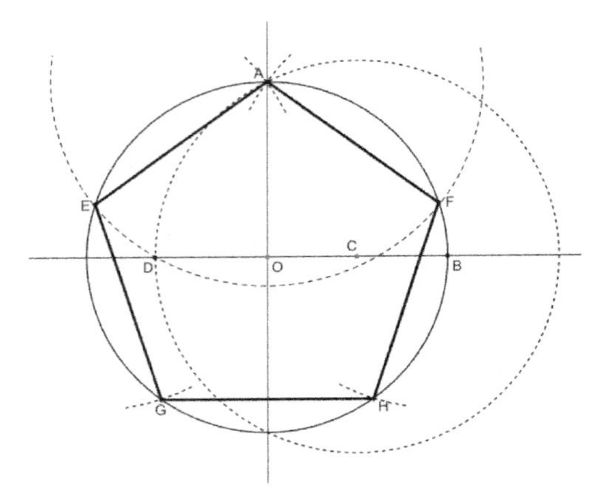

Figure 52. Compass and Straightedge Construction of a Pentagon

Just a small proportion of regular polygons (n-gons) can be constructed with compass and straightedge (Figure 52 shows the construction of the pentagon). Gauss proved that if n is a Fermat prime, then it is possible to construct an n-gon. Wantzel later proved this condition was also necessary (for prime n-gons), so the 65537-gon is currently the largest known constructible prime n-gon. It took Hermes 10 years and a 200-page manuscript to write down a procedure for its construction. Would you like to attempt it?

65539
The largest known Fermat prime + 2, making it the larger member of a twin prime pair.

65837
The largest and only multidigit prime factor in Harold Smith's phone number (4937775). His brother-in-law (Albert Wilansky of Lehigh University) was the first to notice that the sum of its digits is equal to the sum of the digits in its prime factors. Composite integers with this property are called Smith numbers.

66347
The smallest prime that is the sum of the cubes of three consecutive primes: $23^3 + 29^3 + 31^3 = 66347$. [Gallardo]

Figure 53. Classical Construction Problems

Classical geometry uses only a (collapsible) compass and (unmarked) straightedge. On page 166 we show a typical construction; that of a pentagon. Starting with two fixed points a unit apart, what can we construct? Examples include all of the regular n-gons, where n is a power of two times a product of distinct Fermat primes. We can construct an angle of n degrees (n an integer) only if it is divisible by 3. The constructible lengths are called constructible numbers.

Perhaps just as interesting is what cannot be constructed, and how often primes are involved in these problems. Below we give the three most famous examples.

According to a Greek legend, the citizens of Athens consulted the oracle at Delos in order to get rid of a certain pestilence. The oracle required doubling the size (volume) of Apollo's cubical altar, which meant a side had to be increased by a factor of $\sqrt[3]{2}$ (the so-called Delian constant). The Athenians had been given an impossible task—this "doubling the cube" is unsolvable with only compass and straightedge, because $\sqrt[3]{2}$ is not a constructible number!

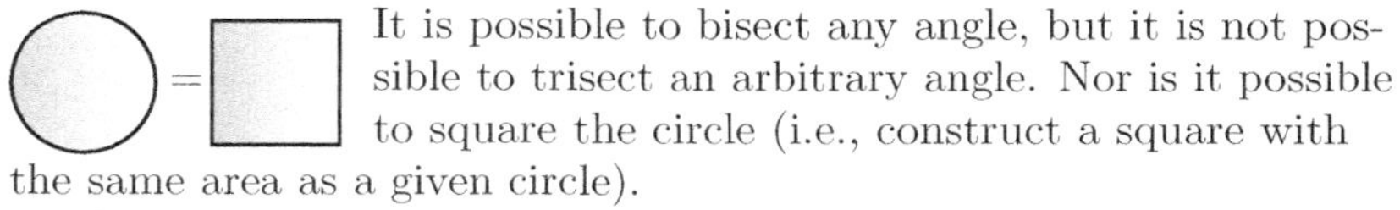

It is possible to bisect any angle, but it is not possible to trisect an arbitrary angle. Nor is it possible to square the circle (i.e., construct a square with the same area as a given circle).

If we alter the rules and allow marks on our straightedge (or use origami instead), then we can construct a larger set of objects including the regular 7-gon, 13-gon, and 19-gon. In fact any p-gon, where p is a prime of the form 2^m3^n+1. Archimedes (circa 287–212 B.C.) showed that we could trisect an angle with these altered rules.

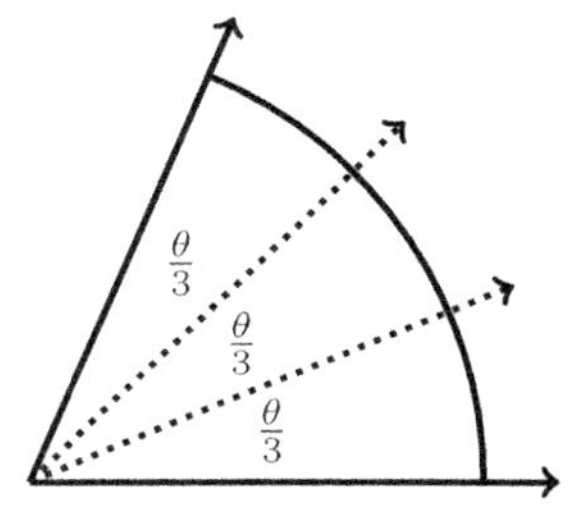

68743
The numbers 68743 + 6!, 68743 + 8!, 68743 + 7!, 68743 + 4!, and 68743 + 3! are all primes. [Honaker]

71317
A palindromic prime that can be expressed as the sum of consecutive primes in three ways. [Rivera]

72169
Humans first stepped onto the Moon on 7/21/69 (Universal Time).

72727
The sum of the digits of the first 72 primes is 727, another palindromic prime. Submitted on 7/27. [Trotter]

73037
The smallest prime that can be represented as the sum of a square and its reversal in two different ways. [Gupta]

76543
The smallest prime number with five consecutive decreasing digits. [De Geest]

77171
$(p_7{}^7 - 1)/(p_7 - 1)$ is prime, where p_7 denotes the 7th prime. [Beedassy]

77269
Subtract 2 from any digit of 77269 and it will remain prime. [Blanchette]

77773
The largest five-digit prime number composed of prime digits. [Hultquist]

78317
In 1867, Swedish chemist, inventor, and philanthropist Alfred Bernhard Nobel received U.S. patent number 78317 for his dynamite. [Blanchette]

78787
A palindromic prime whose reciprocal has a palindromic period length of 39393. [McCranie]

79561

The larger member in the one thousandth twin prime pair.

79999

The largest assigned 5-digit ZIP Code that is a near-repdigit prime. It belongs to El Paso, TX.

80387

The Intel 80387 math coprocessor is a prime example of a Floating Point Unit (FPU).

81457

Leetspeak is a written code or argot (slang) among various Internet communities. The number 81457 spells 'blast' in one common dialect: 8 = b, 1 = l, 4 = a, 5 = s, 7 = t, ... You can have a blast with prime curios! [Earls]

81517

The smallest prime formed from the concatenation (in increasing order) of integral edges of a right triangle (8-15-17). [Patterson]

81619

The largest known prime whose (beastly) square is composed of only two different digits. [Rivera]

81649

The largest prime containing all non-prime digits (from 1 to 9) such that every string of two consecutive digits are squares. [Axoy]

81839

The largest known prime Fibonacci number is fib(81839). It was proven by David Broadhurst and Bouk de Water in 2001.

83431

Retired Japanese engineer Akira Haraguchi once recited π to 83431 decimal places from memory, almost doubling the previous record held by another Japanese. He views the memorization of π as "the religion of the universe." [Haga]

84551

Edward Waring (1736–1798) wrote on algebraic curves, classifying quartic curves into 84551 subdivisions.

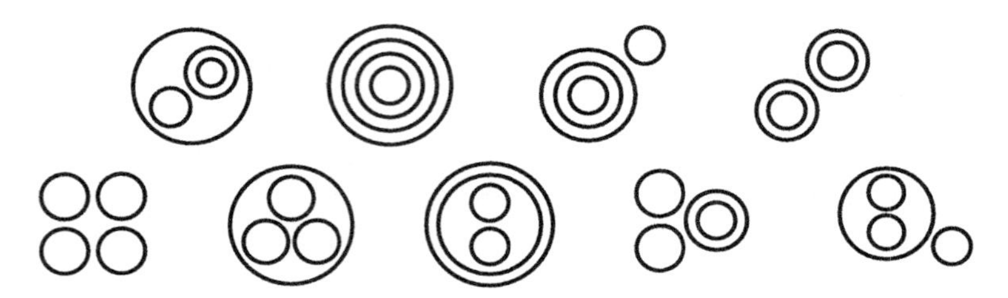

Figure 54. The Nine Ways to Arrange Four Nonoverlapping Circles

84631

The largest known prime such that removing the right-most digit leaves you with a right-truncatable positive integer which is divisible by the sum of its digits in base ten. Such integers are called Harshad (or Niven) numbers. [Earls]

85837

85837 = (8! + 5! + 8! + 3! + 7!) + (8 + 5 + 8 + 3 + 7). It is the only prime number with this property. [Firoozbakht]

86269

Replace any single digit of 86269 with a 3 and the result is always prime. It is the largest five-digit prime with this property. [Luhn]

87811

The number of ways of arranging 14 nonoverlapping circles (see, for example, Figure 54) and also the number of rooted trees with 15 nodes.

90053

Reverse the prime number 90053 and turn it upside down on a calculator to end a "wild GOOSE chase." [Kulsha]

90089

The smallest circular-digit emirp. [Post]

90863

The largest curved-digit prime that contains only distinct curved digits. [Colucci]

92857

The sum of the first n consecutive prime numbers up to 92857 contains all nine digits exactly once (zeroless pandigital).

93701

$\pi(93701) = 9^4 + 3^4 + 7^4 + 0^4 + 1^4$.

94049

The only palindromic prime of form $ab0ba$ in which both ab and ba are composite.

94397

The largest known prime (emirp) that produces all distinct primes by deleting any one digit. [Rivera]

94649

The smallest palindromic prime whose digits are all composite. [Honaker]

95471

The largest prime with distinct digits such that each digit is **coprime** to any of its other digits. [Murthy]

95479

The smaller member in the first pair of Siamese primes that are both emirps. Siamese primes are prime pairs of the form $(n^2 - 2, n^2 + 2)$. The sequence of pairs starts (7, 11), (79, 83), (223, 227), (439, 443), and (1087, 1091). These were named by Beauregard and Suryanarayan in 2001.

95731

The largest prime (emirp) that contains only distinct odd digits. [Poo Sung and Loungrides]

95959

The largest palindromic "prime time" of day on a clock in hours, minutes, and seconds (9:59:59). [Beedassy]

98689

The smallest palindromic circular-digit prime. [De Geest]

The smallest prime that remains prime when inserting one, two, three, four, or five zeros between each digit (908060809, 9008006008009, etc.). [Vrba]

99929

The largest assigned 5-digit ZIP Code that is prime. It belongs to Wrangell, AK.

100019

The smallest six-digit prime whose digits are not prime. [Gevisier]

100057

The first missing prime in the first million digits of π. [Blanchette]

100129

If n is an odd triperfect number (divisors sum to $3n$), then the largest prime factor of n is at least 100129. [Cohen and Hagis]

101723

The smallest prime whose square contains all of the digits from 0 to 9. [Gupta]

101929

A prime formed by combining the first and last three-digit palindromic primes: 101 and 929. [De Geest]

102251

The lesser prime in the smallest set of six consecutive primes whose sum of digits is another set of six distinct primes. [Gupta]

103049

In *Moralia*, the Greek biographer and philosopher Plutarch of Chaeronea states, "Chrysippus says that the number of compound propositions that can be made from only ten simple propositions exceeds a million." The astronomer and mathematician Hipparchus refuted this by showing that on the affirmative side there are 103049 compound statements. [Post]

103993

The first ten digits of 103993 divided by 33102 match the first digits of π.

104869

The smallest prime that contains all of the non-prime digits. [De Geest]

108109

The smallest invertible prime formed by concatenating two consecutive integers: 108 and 109. [Punches]

110237

The record-breaking amount in cash and prizes won by contestant Michael Larson on the TV game show *Press Your Luck* in 1984.

111119

The smallest prime that does not divide at least one ten-digit pandigital number.

111121

A prime formed from the first three terms of the geometric progression $(1, 11, 121)$ whose common ratio is the smallest double-digit prime. [Bhattacharya]

111317

A prime number formed by concatenating the three smallest double-digit primes $(11, 13, 17)$ in order. Note that if you concatenate the 11th prime, the 13th prime, and the 17th prime, in order, you'll get another prime (one whose digits begin a well-known mathematical constant). [Somer]

112643

The ill-fated Feit-Thompson conjecture claimed that there were no primes p and q for which $\frac{p^q-1}{p-1}$ and $\frac{q^p-1}{q-1}$ have a common factor. In 1971, N. M. Stephens found 112643 divides both $\frac{17^{3313}-1}{17-1}$ and $\frac{3313^{17}-1}{3313-1}$. (A second example is hard to come by!) [Post]

113797

An emirp formed from two three-digit numbers, 113 and 797, that are each reversible primes. [Card]

114593

The smallest prime formed from the first six digits of the decimal expansion of π, i.e., 314159.

116911

The smallest strobogrammatic emirp. [Punches]

120121
The smallest **dihedral prime** in which all four associated primes are distinct. [Keith]

121661
The 121st prime is 661. If we concatenate 121 and 661, or 661 and 121, we get two more primes (121661 and 661121). This is the smallest case of what high school algebra teacher Terry Trotter (1941–2004) called OP-PO primes (O = Ordinal position, and P = Prime).

123457
When you remove the penultimate (next to the last) digit of 123457 you will end up with yet another prime. Repeat the process for four more primes. [Brod]

125959
The largest "prime time" of day on a 12-hour clock in hours, minutes, and seconds (12:59:59). [Punches]

131143
A prime number that can be written as $2^{17} + 71$. (17 and 71 are emirps.)

135799
The smallest prime that contains all of the odd digits in increasing order. [Gupta]

138239
The number of different unsolved configurations that can be reached on the Skewb Diamond. The Skewb Diamond is an octahedron-shaped puzzle similar to the Rubik's Cube. [Rupinski]

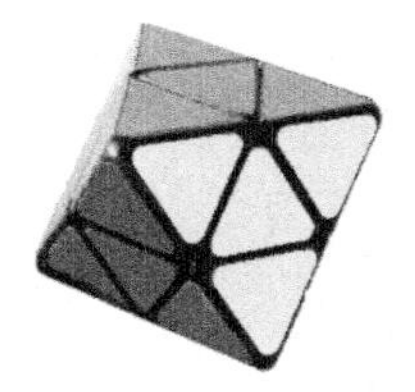

139967
The lesser prime in the smallest twin prime pair for which the sum is a seventh power: 6^7. Note that this number ends with 6 and 7. [Rivera and Trotter]

174763
One of the prime factors of $2^{95} + 1$. Found with D. H. Lehmer's Photoelectric Sieve (see page 67) in the 1930's. [Haga]

180001
The sum of all digits of the first ten thousand positive integers. [Beedassy]

185527
The smallest prime expressible as the sum of three distinct prime cubes in more than one way: $185527 = 19^3+31^3+53^3 = 13^3+43^3+47^3$. [Beedassy]

206483
The smallest prime that contains all of the even digits. [Gupta]

215443
The first odd prime formed from the leading digits of the decimal expansion of $\sqrt[3]{10}$. [Gupta]

220373
The minimum number of colors necessary to color any map drawn on a stage-8 Menger sponge (the unit cube is stage-0) so that no regions sharing a common border receive the same color. [Post]

235813
The smallest prime formed from the concatenation of five consecutive Fibonacci numbers. [Gupta]

235951
The largest "prime time" of day on a 24-hour clock in hours, minutes, and seconds (23:59:51). [Patterson]

238591
The number of triskaidecominoes (13-polyominoes). [Hartley]

241603
The start of thirteen consecutive primes each congruent to 3 (mod 4). A sprint in the prime number race (page 159)!

246683
Richard G. E. Pinch found that there are exactly 246683 Carmichael numbers below 10^{16}.

253987
The largest known "near-square" prime of the form $(2^k - 1)^2 - 2$ has index 253987. [Emmanuel]

265729

A Delannoy number $D(n,n)$ is the number of lattice paths from $(0,0)$ to (n,n) in an n-by-n grid using only steps North, Northeast, and East (Figure 55). There are only three known that are prime: 3, 13, and 265729 (corresponding to indices $n = 1, 2$, and 8). The next one must have over 300 digits! [Post]

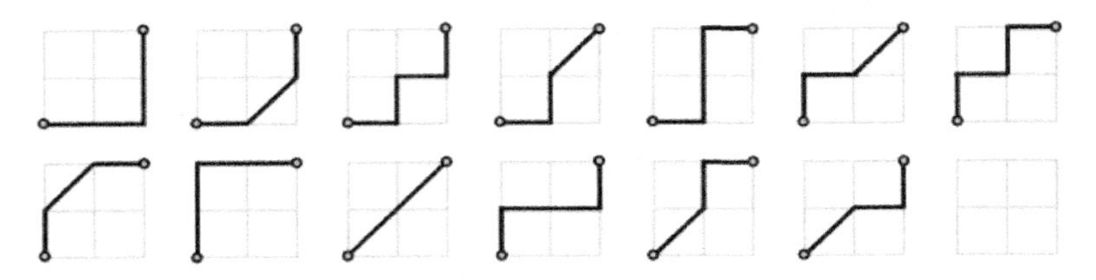

Figure 55. The 13 Paths for the Second Central Delannoy Prime

271129

The smallest known prime Sierpiński number (2nd kind): $271129 \cdot 2^n + 1$ is composite for every positive integer n. [Rupinski]

274177

The smallest factor of the second composite Fermat number. Retired mathematician Fortune Landry discovered it in 1880, at the age of 82.

281581

The long-standing record number of dominoes set up and toppled by a single person. Klaus Friedrich achieved the feat in 1984. [Rupinski]

292319

The smallest prime formed from the reverse concatenation of three consecutive primes. [Gupta]

294001

The smallest **weakly prime** (changing any of its digits produces a composite). [Weisstein]

304807

The number of characters in the Koren edition of the Torah including two instances where the letter nun is written upside down. Note that 304807 is an emirp. [Croll]

314159

A prime number embedded in the decimal expansion of π. [Gardner]

The six-digit combination to Ellie's small office safe in the novel *Contact* by Carl Sagan.

331997

$331997 = 1!^1 + 2!^2 + 3!^3 + 4!^4$ is prime. [Poo Sung]

355777

A prime number formed by the concatenation of one copy of the first odd prime, two copies of the second odd prime, and three copies of the third odd prime. [Grant]

369119

A prime number that divides the sum of all primes less than or equal to 369119. [Nelson]

369293

The smallest prime that cannot become another by inserting a digit. [Blanchette]

379009

A prime number that spells *Google* when turned upside down on a calculator. [Earls]

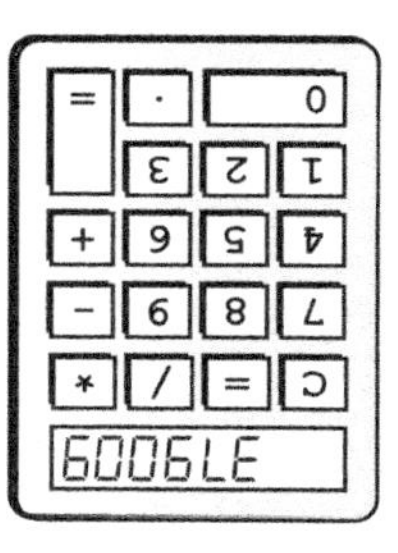

396733

The first occurrence of a prime in which the next consecutive prime is found by adding 100. [Glaisher]

400031

Gauss used Vega's table of primes up to 400031 to further corroborate his belief that the distribution of primes is inversely proportional to the logarithm (see *prime number theorem* on page 261).

467479

A concatenation of two consecutive primes that forms the last member of a prime quadruple.

472393

If a number can be written both as $2a^2 + 1$ and as $3b^3 + 1$, it must be of the form $648k^6 + 1$. The smallest prime of this form is 472393, where $k = 3$. (It is also prime for $k = 4, 5, 8, 13, 16, 19, 20, 24, \ldots$.) [Hartley]

509203

It is conjectured that 509203 is the smallest Riesel number, i.e., the smallest integer k such that $k \cdot 2^n - 1$ is composite for all $n \geq 0$.

513239

The greatest prime factor of any eleven-digit repdigit. [La Haye]

514229

A Fibonacci prime and the 29th Fibonacci number. Note that it ends with 29. [Axoy]

539633

The smallest prime number of the form $a^m + b^m + c^m + \ldots$, where $a \cdot b \cdot c \cdot \ldots$ is the prime factorization of m. [Honaker]

The sum of 7 positive 11th powers. [Sloane]

641491

The prime $p = 641491$ is the smallest known counterexample to Masley's suggestion of how one might prove Vandiver's conjecture, i.e., show that the class number $h(Q(\zeta_p + \zeta_p^{-1}))$ is smaller than p. [Poo Sung]

649739

Odds against drawing a royal flush in poker: 649739 to 1. [Rogowski]

664579

The number of prime numbers less than ten million (see Table 10). [Dobb]

680189

The smallest **prime rotative twin** (invertible prime) to contain five different digits. [Trigg]

686989

The smallest composite-digit strobogrammatic prime. [Beedassy]

Table 10. The Number of Primes Less Than x

x	$\pi(x)$
10	4
100	25
1,000	168
10,000	1,229
100,000	9,592
1,000,000	78,498
10,000,000	664,579
100,000,000	5,761,455
1,000,000,000	50,847,534
10,000,000,000	455,052,511
100,000,000,000	4,118,054,813
1,000,000,000,000	37,607,912,018
10,000,000,000,000	346,065,536,839
100,000,000,000,000	3,204,941,750,802
1,000,000,000,000,000	29,844,570,422,669
10,000,000,000,000,000	279,238,341,033,925
100,000,000,000,000,000	2,623,557,157,654,233
1,000,000,000,000,000,000	24,739,954,287,740,860
10,000,000,000,000,000,000	234,057,667,276,344,607
100,000,000,000,000,000,000	2,220,819,602,560,918,840
1,000,000,000,000,000,000,000	21,127,269,486,018,731,928
10,000,000,000,000,000,000,000	201,467,286,689,315,906,290
100,000,000,000,000,000,000,000	1,925,320,391,606,803,968,923

728729

The smallest prime formed by concatenating consecutive Smith numbers (Smith brothers). [Necula]

739397

The largest two-sided prime (both right and left-truncatable).

767857

Add 1 to any digit of 767857 except the last one and it will remain prime. [Blanchette]

776887

The smallest emirp of the form $n^n - (n-1)^{n-1}$.

797161

$797161 = \frac{3^{13}-1}{3-1}$. [Sylvester]

800801

The reflectable prime number of years traveled by the time traveler in H. G. Wells' *The Time Machine* upon his return from the year 802701 to 1900. [Rupinski]

823541

The smallest odd prime that may be expressed as $p^p - 2$, for some prime p. [Gallardo]

823679

The only secret access number that John Nash saw on his forearm in the movie *A Beautiful Mind* (Universal Pictures, 2002). [Rupinski]

823799

The largest prime of the form $a^a + b^b$, where a and b are one-digit positive integers. [Gupta]

864203

The smallest prime that contains all of the even digits in decreasing order. [Gupta]

908909

A circular-digit prime composed of two consecutive numbers (908 and 909). [Punches]

909287

The smallest prime in the first occurrence of five consecutive sets of twin primes. [Dobb]

919799

The largest number of votes ever cast for a U.S. presidential candidate who was imprisoned at the time of the election (Eugene V. Debs, Socialist, 1920). [NYU Law Review]

938351

Subtract 1 from any digit of 938351 (except the 1) and it will remain prime. [Blanchette]

953593

The number of particles that compose the disintegrating chair that Po blows up with fireworks in the computer-animated film *Kung Fu Panda* (from a Hewlett-Packard Company fact sheet). [Capelle]

975313

The smallest prime that contains all of the odd digits in decreasing order. [Gupta]

989999

The 77777th prime number. [Edwards]

999331

The largest known circular prime that is not a repunit prime.

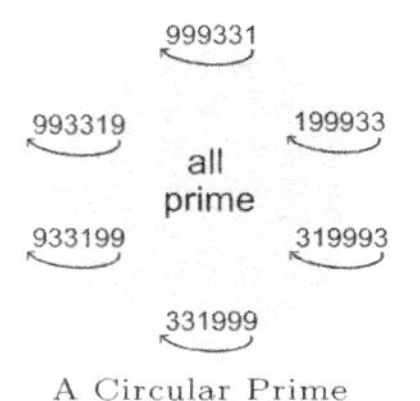

A Circular Prime

999983

The largest six-digit prime (emirp). [Gallardo]

1000003

The smallest prime greater than a million. [Kumar]

1002887

The first "missing prime" in the first one hundred million digits of π. [De Geest]

1023467

The smallest prime to contain seven distinct digits. [Edwards]

1023487

The smallest emirp to contain seven distinct digits. [Post]

1062881

The natural numbers can be arranged into an infinite list of ratios, all of which are integers:

$$(1+2)/3,$$
$$(4+5+6+7+8+9+10+11)/12,$$
$$(13+14+15+16+\ldots+35+36+37+38)/39,$$
$$(40+41+42+43+44+\ldots+115+116+117+118+119)/120.$$

The thirteenth of these ratios is the prime 1062881.

1068901
The smallest strobogrammatic prime to contain five different digits. [Trigg]

1201021
The smallest seven-digit palindromic prime with a sum of digits equal to seven. [Murthy]

1203793
The largest known **Gaussian Mersenne prime** is $(1+i)^{1203793} - 1$ (found September 2006 by B. Jaworski). The real and imaginary parts of this behemoth both have 180989 digits and its norm (the square root of its absolute value) is a 362378-digit prime. Figure 56 shows how the smaller Gaussian primes are distributed. There are 36 smaller Gaussian Mersennes known.

1212121
The smallest seven-digit smoothly undulating palindromic prime number.

1237547
The smallest seven-digit left-truncatable prime. [Opao]

1264061
$\pi(1264061) = 1^6 + 2^6 + 6^6 + 4^6 + 0^6 + 6^6 + 1^6$.

1265347
The smallest emirp to contain the distinct digits 1 through 7. [Punches]

1304539
The mean prime gap up to 1304539 is exactly one dozen (see Table 11 on page 184). [Dodson]

1370471
Prime Period Lengths by Samuel Yates contains the period lengths of all primes (except 2 and 5, whose reciprocals are terminating decimals) up to 1370471.

1409041
The smallest palindromic prime that contains all of the square digits. [Gupta]

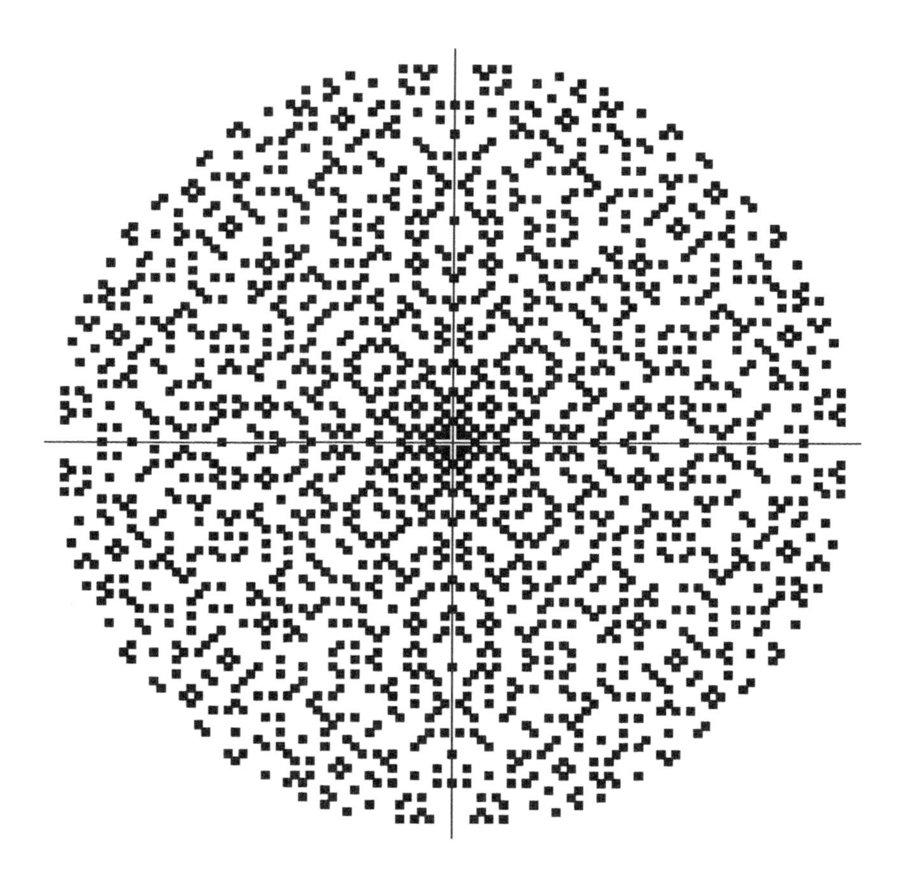

Figure 56. Gaussian Primes

The Small Gaussian Primes. Gaussian primes are the primes in the set of numbers of the form $a + bi$, where a and b are integers. The center of the image above is 0, and the primes $a + bi$ are graphed at the coordinates (a, b). Gaussian primes come in two flavors: the primes $\pm a$ and $\pm ai$, where a is a real prime congruent to 3 modulo 4; and $a \pm bi$, where $a^2 + b^2$ is a real prime. Terence Tao has proven that all possible constellation shapes appear in these primes!

Table 11. First Occurrences of Mean Prime Gaps

n	nth prime	mean gap
2	3	0
10	29	2
68	337	4
438	3,061	6
2,616	23,537	8
100,350	1,304,539	12
637,198	9,557,957	14
27,066,970	514,272,413	18
179,992,840	3,779,849,621	20
55,762,149,072	1,505,578,024,919	26
382,465,573,492	11,091,501,631,241	28
2,636,913,002,950	81,744,303,091,421	30
126,979,448,983,454	4,444,280,714,420,857	34
885,992,692,751,466	32,781,729,631,804,207	36
43,525,513,764,814,972	1,784,546,064,357,413,813	40
306,268,030,480,171,300	13,169,525,310,647,365,859	42

1428571

The first seven digits of the decimal expansion of the reciprocal of seven. [Glenn]

1490941

The smallest palindromic prime of the form "prime0emirp." Note that all the digits are squares. [Beedassy]

1492637

An emirp that can be expressed as the sum of squares of five consecutive primes. [Patterson]

1618033

A prime number formed from the first seven digits of the golden ratio: $\phi = \frac{1+\sqrt{5}}{2} = 1.618033....$ [Gupta]

1676267

The number of inequivalent Latin squares of order eight. Two squares are equivalent (or isotopic) if one can be obtained from the other by permuting rows, columns, and symbols.

1705829

The first of fifty-seven distinct primes generated by the absolute value of

$$(n^5 - 133n^4 + 6729n^3 - 158379n^2 + 1720294n - 6823316)/4,$$

where $n = 0$ to 56. Shyam Sunder Gupta of India is credited with this discovery.

1723609

Thomas A. Edison received U.S. patent number 1723609 for an "Apparatus for Producing Storage-Battery Electrode Elements." It is the largest patent number he held that was prime.

1777541

Bloomsbury Publishing Plc announced that a record-breaking 1,777,541 copies of *Harry Potter and the Order of the Phoenix* were sold on the first day of its release (June 21, 2003).

1820281

A palindromic prime that is one more than the sum of the twin primes 910139 and 910141. (1820281 is also a twin prime!) [Vouzaxakis]

1823281

Did you know that 1823 is the 281st prime? Together they become a palindromic prime. [De Geest and Rosa]

1831381

A **palindromic reflectable prime**. [Trigg]

1865491

The least prime in a chain of three consecutive primes such that all three are emirps and have the same sum of digits. [Earls]

1880881

The smaller prime in the first tetradic Gridgeman pair (see 181 on page 75). [Punches]

1934063

The first of two consecutive prime numbers whose product is palindromic. [De Geest]

Table 12. Sets of n Primes with Prime Subset Means (with the Smallest Possible Largest Element)

n	"Averaging" Set of n Primes
2	$\{3, 7\}$
3	$\{7, 19, 67\}$
4	$\{5, 17, 89, 1277\}$
5	$\{209173, 322573, 536773, 1217893, 2484733\}$

2024689
The smallest prime that contains all of the even digits in increasing order. [Gupta]

2124679
The largest known Wolstenholme prime.

2222243
The first seven-digit prime that has only one odd digit. [Fentsor]

2222273
The smallest prime formed by the concatenation of seven primes. [Colucci]

2442113
The start of fifteen consecutive prime numbers whose sum of digits is prime.

2484733
The means of each of the 31 nonempty subsets of {209173, 322573, 536773, 1217893, 2484733} are distinct prime numbers. Sets for which every nonempty subset has a prime mean are called averaging sets of primes. The smallest of these (those with the least possible largest element) are in Table 12. [Granville]

2696063
$\frac{1}{2} - \frac{1}{3} + \frac{1}{5} - \frac{1}{7} + \frac{1}{11} - \frac{1}{13} + \ldots = 0.2696063\ldots.$

2718281
A prime formed from the first seven digits of the decimal expansion of e. [Gupta]

2759677

The largest known prime containing a prime number of digits: $27653 \cdot 2^{9167433} + 1$, with 2759677 digits.

2873113

The sum of all primes of the form $4n + 3$ that are less than ten thousand. [Patterson]

2935241

If $p(n)$ is the nth prime, then $p(p(p(2 \cdot 9 \cdot 3 \cdot 5 \cdot 2 \cdot 4 \cdot 1))) = 2935241$. [Firoozbakht]

2951413

A prime whose reversal begins the decimal expansion of π. [Gardner]

3187813

A palindromic prime equal to the sum of squares of two consecutive integers: $1262^2 + 1263^2$. [De Geest]

3370501

On the first page of his algebraic number theory text, K. B. Stolarsky said that to solve $y^2 + 999 = x^3$ in positive integers is a problem one which "a fool can ask, but a thousand wise men cannot answer." There are just six solutions and 3370501 is the largest of the y values. [Beedassy]

3511973

The lesser prime in the smallest set of seven consecutive primes for which the sum of digits of these primes is another set of seven distinct primes. [Gupta]

3541541

The first prototype computer mouse (1964) was made by Douglas Engelbart. (Computer Mouse U.S. Patent Number 3541541, November 1970). [Bellis]

4405829

The patent number on the **RSA algorithm**—perhaps the most famous of all public-key cryptosystems.

4695947

The sum of the cubes of all the two-digit primes. [Silva]

4784009

The largest possible prime value resulting from the ninedigital fraction equation $(\frac{A}{B})^C + (\frac{D}{E})^F + (\frac{G}{H})^I$, where the letters A to I represent any combination of the nine digits 1 through 9. Note that $(\frac{6}{3})^4 + (\frac{8}{2})^5 + (\frac{9}{1})^7$ is the only distinct solution for 4784009. [De Geest]

4833869

The sum in the following cryptarithm created by Mike Keith:

$$\text{STUFFED} + \text{TURKEY} = \text{DESSERT}.$$

Would you rather see "prime rib" on the mathematical menu?

4973029

The smallest "super-7 prime," i.e., a prime p such that $7p^7$ contains 7 consecutive 7's in its decimal representation. [De Geest]

5009317

Patent 5009317 was issued on April 23, 1991, for a better mousetrap. [NYU Law Review]

5195977

The sum $\frac{1}{2} + \frac{1}{3} + \frac{1}{5} + \ldots + \frac{1}{5195969} + \frac{1}{5195977}$ just exceeds the first odd prime number, or 3. [Schimke]

5224903

$\pi(5224903) = 5! + 2! + 2! + 4! + 9! + 0! + 3!$.

5491763

The largest prime number with distinct digits in English alphabetical order (Five, Four, Nine, One, Seven, Six, Three). [Honaker]

5550113

The home phone number is given as 555-0113 in one episode of the animated TV series *The Simpsons*. [Rupinski]

5780507

$5780507 = 3^4 + 5^6 + 7^8$. [La Haye]

5858581

The smallest prime such that the cube of the sum of its digits equals the product of its digits. [De Geest]

5882353

$5882353 = 588^2 + 2353^2$. [Madachy]

6608099

The smallest strobogrammatic circular-digit prime. [Punches]

6773999

The smallest nonnegative number not expressible as the determinant of a 5-by-5 matrix with elements $1, 2, 3, \ldots, 5^2$. [Pfoertner]

7352537

The smallest palindromic prime using all prime digits. [Murthy]

7469789

The smaller member of the first pair of consecutive primes that are both weakly primes.

7693967

The only seven-digit palindromic prime that is also a left-truncatable prime. [Patterson]

7715177

This palindromic prime may be written as $\frac{347182965}{3+4+7+1+8+2+9+6+5}$. Note that 347182965 uses each of the nonzero digits once, and only once. [Dobb]

7996369

The largest seven-digit emirp whose reversal is a left-truncatable prime. [Opao]

8396981

Sixty-four to the power 8396981 begins with the digits 8396981. [Hartley]

8627527

The smallest prime number that can be expressed in three different ways as the sum of the cubes of three prime numbers: $19^3 + 151^3 + 173^3$, $23^3 + 139^3 + 181^3$, and $71^3 + 73^3 + 199^3$. [Goelz]

8675309

Jenny's telephone number in the hit song "8675309/Jenny" by Tommy Tutone. [Adams]

9136319
In the decimal expansion of π, starting 9128219 digits after the decimal point, you find the digits 9136319. These numbers, 9128219 and 9136319, are consecutive palindromic primes. [De Geest]

9875321
The largest emirp with strictly decreasing digits. [Chandler]

9876413
The largest prime with a prime number of distinct digits. [Bottomley]

9884999
The number of 1's needed to represent all numbers from one to one million in binary. [Vrba]

10006721
In 1914, D. N. Lehmer published the first widely-available lengthy table of primes. His went from 1 (which he considered prime!) up to 10006721.

10010101
A "byte" that is prime in both decimal and binary. Note that the 0's are in prime positions and the 1's are in non-prime positions when read left-to-right. [Honaker]

10153331
The smallest **self-descriptive prime**. The digits of such numbers are described (i.e., one 0, one 5, three 3's, three 1's) in any order. [McCranie]

10235647
The smallest eight-digit prime containing each of the digits zero through seven. [Baker]

10235749
The smallest eight-digit emirp with distinct digits. [Post]

10939771
According to the National Statistical Service of Greece, this was the population of Greece at the beginning of the new millennium. [Saridis]

11111117

Seven 1's prior to the digit 7 is an emirp. [Trotter]

11117123

The smallest member of the first Ormiston triple. [Wilson]

11281811

The first public performance of Beethoven's piano concerto *Emperor* was most likely that of Friedrich Schneider on 11/28/1811 at a concert in Leipzig. [Necula]

11592961

The smallest prime that can be represented as sum of a cube and its reversal in two different ways. [Gupta]

11698691

The smallest zeroless bemirp. [Wilson]

11718829

The largest number of the one thousandth prime quadruple.

11917049

Replace any single digit of 11917049 with a 3 and the result is always prime. It is the only eight-digit prime with this property. [Luhn]

12356789

The smallest prime with the most distinct nonzero digits. [Murthy]

12422153

Replacing each digit d with d copies of the digit d produces another prime throughout three transformations. Confirmed by P. Jobling to be the smallest prime of this type. [Rivera]

13112221

The largest known prime in the look-and-say sequence (starting with 1). To generate a member of the sequence from the previous term, "look" and "say" the digits of the previous term, counting the number of digits in groups of the same digit. For example, 111221 has three 1's, two 2's, and then one 1; therefore 111221 is followed by 312211. The sequence begins $1, 11, 21, 1211, 111221, \ldots$. [Coveiro]

13466917

In 2001, for the first time in recorded prime history, a prime was found with more digits than the square of the year in which it was found: $2^{13466917} - 1$. [Luhn]

Mersenne Stamp

15365639

In 1769, Euler conjectured that the sum of three fourth powers is never a fourth power. However in 1988, Noam Elkies at Harvard discovered a counterexample: $2682440^4 + \mathbf{15365639}^4 + 18796760^4 = 20615673^4$. [Beedassy]

17880419

The start of the smallest Cunningham chain (1st kind) containing four emirps. [Vrba]

18518809

In 1778, Euler proved that 18518809 was prime by showing that it had a single representation in the form $1848x^2 + y^2$. A remarkable feat for his time.

19141939

Both 19141939 and 1914 + 1939 are primes. Note that 1914 and 1939 are the years that WWI and WWII began. [Heller]

19999999

The smallest prime whose sum of digits is a sixth power. [Gupta]

21322319

The smallest autobiographical prime. The digits of such numbers are described (i.e., two 1's, three 2's, two 3's, one 9) in increasing order. [Kapur]

23010067

In October 2007, 23010067 became the largest known prime for which the partition number $p(23010067)$ is also prime. (The partition number $p(n)$ is the number of ways of writing n as a sum of positive integers. So $p(5) = 7$ because 5 is 5, $4+1$, $3+2$, $3+1+1$, $2+2+1$, $2+1+1+1$, and $1+1+1+1+1$.)

23252729

The smallest prime formed from the concatenation of four consecutive odd numbers. [Gupta]

23456789

The largest prime with strictly increasing digits. [Chandler]

25935017

$1!^2 + 2!^2 + 3!^2 + 4!^2 + 5!^2 + 6!^2 + 7!^2$. [Post]

26170819

The discovery that $M(21701)$ is prime occurred exactly 26,170,819 seconds into the year 1978 (Pacific Standard Time). [Noll]

27177289

The last 8 digits of 9^{9^9}. [Bakst]

29061379

The last time four consecutive days were each prime in the *dd/mm/yyyy* format began on 29/06/1379. This was the day after Petrus Boeri was deprived of his bishopric by Pope Urban VI, for supporting Clement VII. [Hartley]

29113327

The next time four consecutive days are each prime in the *dd/mm/yyyy* format begins on 29/11/3327. [Hartley]

31114073

Replace any single digit of 31114073 with a 9 and the result is always prime. [Luhn]

35009333

$1 \cdot 1^1 + 11 \cdot 2^2 + 111 \cdot 3^3 + 1111 \cdot 4^4 + 11111 \cdot 5^5$. [Earls]

35200001

The smallest prime p such that the digit 1 appears exactly p times among all positive integers not exceeding p. [Beedassy]

39916801

The smallest factorial prime that is also an emirp. [Loungrides]

45269999

$9^8 + 8^7 + 7^6 + 6^5 + 5^4 + 4^3 + 3^2 + 2^1 + 1^0$. [Kulsha]

48205429

The smallest integer that can be expressed as the sum of consecutive primes in exactly 9 ways. [Meyrignac]

50943779
The smallest prime p such that $n \pm 16, n \pm 8, n \pm 4, n \pm 2$ are each prime for $n = p + 16$. [Ellermann]

53781811
The answer to how many complete copies of the Gutenberg Bible on vellum are known to exist can be found by reading 53781811 turned upside down on a calculator.

56598313
The largest known prime p for which p^2 divides $10^p - 10$. Note that 3 and 487 are the only smaller values. [Richter]

59575553
The smallest prime formed from the reverse concatenation of four consecutive odd numbers. [Gupta]

60000607
The absolute values of -60000607, $6 - 0000607$, $60 - 000607$, $600 - 00607$, $6000 - 0607$, $60000 - 607$, $600006 - 07$, and $6000060 - 7$ are each prime. There is no larger example known. [Rivera]

61277761
The reversal of 61277761 is 8^8. There should be infinitely many primes of this form. Can you find another? [Kulsha]

66600049
The largest minimal prime in base ten. Choose any prime number, and by deleting zero or more of its digits, you can arrive at one of the following twenty-six minimal primes:

> 2, 3, 5, 7, 11, 19, 41, 61, 89, 409, 449, 499, 881, 991, 6469, 6949, 9001, 9049, 9649, 9949, 60649, 666649, 946669, 60000049, 66000049, 66600049.

These were first defined (and discovered) by J. Shallit. [Rupinski]

67867967
This easy-to-remember number is the four millionth prime. [Edwards]

67898771
The largest number not expressible as a sum of distinct fifth powers. [Beedassy]

73939133
The largest right-truncatable prime.

76540231
The largest eight-digit prime containing each of the digits 0 through 7. [Baker]

77345993
An "egg-cellent" prime! Note that if you type it into a calculator, then turn the calculator upside down, it looks as if the word `EGGShELL` is spelled out. [Earls]

91528739
The least of ten consecutive emirps. [Russo]

92525533
The smallest emirp formed by concatenating a primitive Pythagorean triple ($92^2 + 525^2 = 533^2$). [Post]

98303927
The smallest prime that is the average of the eight surrounding primes (four on each side). [McCranie]

98765431
The largest prime with strictly decreasing digits. [Chandler]

100000007
The largest prime factor of an odd perfect number must be at least 100000007. [Goto and Ohno]

100330201
Yakov Kulik (1783–1863) is said to have spent two decades constructing a table of factors of numbers less than 100330201. Volume 2 (of 8) is now missing.

102345689
The smallest prime with the most distinct digits. [Papazacharias]

102346897
The smallest emirp with the most distinct digits. [Beedassy]

104395289

In 1959, C. L. Baker and F. J. Gruenberger prepared a table of prime numbers up to 104395289 that were arranged on microcards. Each of the 124 photographs (except the first and last) displayed 39 pages of 1250 primes.

111113111

Replacing each digit d with d copies of the digit d produces another prime through two transformations. [Keith]

111181111

The smallest square-congruent prime.

1	1	1
1	8	1
1	1	1

122333221

A palindromic prime that begins with one 1, two 2's, and three 3's, when read left-to-right or right-to-left. [Murthy]

123575321

A palindromic prime whose digits are ordered consecutive primes if 1 is accepted as prime. [Trigg]

135979531

The smallest palindromic prime containing all of the odd digits. [Gupta]

138140141

The smallest prime formed from the concatenation of three consecutive composite numbers. [Gupta]

172909271

The smallest palindromic prime that contains the famous Hardy-Ramanujan number 1729 (page 28). [Gupta]

186690061

The smallest prime that when turned upside down, becomes a product of two tetradic primes. [Blanchette]

193707721

Lucas demonstrated that $2^{67} - 1$ (which Mersenne had claimed to be prime) was composite in 1876; but he did so without finding its factors. In a 1903 meeting of the American Mathematical Society, F. N. Cole (1861–1926) silently walked to the board and calculated

$2^{67} - 1$. On another board he multiplied the primes 193707721 and 761838257287, and still in silence, sat down to a standing ovation.

197072563

$4p^2 + 1$ is prime for $p = 197072563$. The new prime will produce yet another prime if placed back into the original formula. This iteration can be repeated for a total of 4 new primes. [Rivera]

207622273

Another prime race sprint: this is the first of sixteen consecutive $4n + 1$ primes. [Vandemergel]

222222227

The smallest nine-digit **prime-digit prime**. [Murthy]

261305843

The smallest prime number to be the sum of the cubes of the first consecutive odd primes $(3^3 + 5^3 + 7^3 + \ldots + 271^3)$. [Capelle]

285646799

$2 \cdot 19 \cdot 23 \cdot 317 \cdot 1031 + 1$ is prime. Note the five known repunit prime digit lengths.

289327979

The smallest prime such that adding two preceding or trailing digits (not simultaneously equal to zero) will always result in a composite number. [Noll]

298995971

The last member of the only known prime quintuplet (3, 11, 131, 17291, 298995971) in a $p^2 + p - 1$ progression. [De Geest]

303272303

The smallest palindromic prime p such that $1p1$, $3p3$, $7p7$, and $9p9$, are all primes. [De Geest]

304589267

Arrange the nine distinct digit prime 304589267 as $\frac{30}{45} + \frac{89}{267}$ to reveal the "1" missing digit. There is only one other prime for which this is possible. [Trotter and Knop]

Table 13. Least prime p such that 2^p-1 has 10^n+ digits

n	prime p	n	prime p
0	2	12	3321928094941
1	31	13	33219280948907
2	331	14	332192809488739
3	3319	15	3321928094887411
4	33223	16	33219280948873687
5	332191	17	332192809488736253
6	3321937	18	3321928094887362349
7	33219281	19	33219280948873623521
8	332192831	20	332192809488736234933
9	3321928097	21	3321928094887362347897
10	33219280951	22	33219280948873623478723
11	332192809589	23	332192809488736234787093

Note: $\frac{\log(10)}{\log(2)} = 3.32192809488736234787031942948939017...$

306989603
The smallest palindromic prime that contains all of the curved digits. [Gupta]

313713137
With 313713139 forms a pair of reversible twin primes. [Card]

317130731
The start of the smallest set of five consecutive prime numbers such that each term is the sum of the previous term plus its sum of digits. [Rivera]

332192831
The smallest Mersenne number with a prime exponent that contains 100 million or greater decimal digits is $M(332192831)$ (see Table 13). [Firoozbakht]

333667001
The smallest prime that is equal to the sum of the cubes of its third parts: $333^3 + 667^3 + 001^3$. [Rivera]

345676543

Former editor Léo Sauvé once pointed out the particularly attractive nine-digit palindromic prime 345676543 in an issue of the problem solving journal *Crux Mathematicorum.*

352272253

The smallest palindromic prime of the form "primemirp" containing all of the prime digits. [Beedassy]

354963229

The smallest number in a set of six consecutive primes whose sum of digits are prime and equal. [Gupta]

363818363

A palindromic prime whose reciprocal has a palindromic period length. [McCranie]

373587883

The twenty-millionth prime number is the concatenation of three 3-digit primes: 373, 587, and 883. [Dobb]

375656573

The smallest palindromic prime that is also a Sierpiński number (page 23). [Rivera]

377333773

The only nine-digit palindromic prime composed of threes and sevens alone. [Trigg]

378163771

The prime number 378163771 is ILLEgIBLE if turned upside down on a calculator. [Keith]

387947779

The ISBN (International Standard Book Number) for the first edition of Crandall and Pomerance's excellent book *Prime Numbers: A Computational Perspective.* [Mihai]

415074643

The 22096548th prime number, 415074643, divides the sum of the first 22096548 primes. There are no larger examples less than 10^{12}. [Chua]

433494437
The first Fibonacci number that contains its index twice is prime. [Axoy]

446653271
When 446653271 is squared, each digit of the resulting number is a square. [Dobb]

452942827
The first of eleven consecutive primes that end in seven. [Rupinski]

486894913
Insert a 3 anywhere to form $3 \cdot 3$ other primes. [Blanchette]

506977979
$\sqrt{5! + 0! + 6! + 9! + 7! + 7! + 9! + 7! + 9!}$ is prime.

527737957
$5\underbrace{27737957}$ is the $\underbrace{27737957}$th prime number. [Blanchette]

529510939
One of the prime factors of $2^{93} + 1$. Found with D. H. Lehmer's Photoelectric Sieve (see page 67) in the 1930's. [Haga]

540298673
The largest prime number with distinct digits in Spanish alphabetical order (Cinco, Cuatro, Cero, Dos, Nueve, Ocho, Seis, Siete, Tres). [Rivera]

587523659
A "musical" prime obtained by concatenating the rounded frequencies of the musical notes D_5 (587 Hz), C_5 (523 Hz), and E_5 (659 Hz). Each of the frequencies is prime too. [Necula]

593103437
Insert a 9 anywhere to form 9 other primes. [Blanchette]

608844043
The values of $+608844043$, $6 + 08844043$, $60 + 8844043$, $608 + 844043$, $6088 + 44043$, $60884 + 4043$, $608844 + 043$, $6088440 + 43$, and $60884404 + 3$ are primes. No larger prime with this property is known. [Rivera]

608888809

The smallest strobogrammatic prime containing all 8's in between 60 and 09. [Krussow]

627626947

The largest nine-digit left-truncatable prime such that when two of its digits are interchanged, the resulting number is also a left-truncatable prime. Can you locate the two digits in question? [Opao]

684972991

The smallest prime having over one million prime prime-residues. The expression "prime prime-residues" comes from a puzzle by Rickey Bowers, Jr. When you divide 17 (for example) by the primes less than it, you get the remainders 1, 2, 2, 3, 6, and 4. Three of these remainders (residues) are prime. So he said that 17 has three prime prime-residues. [Oakes]

723121327

Choose any digit *d* of this number; there is always a matching *d* spaced *d* digits apart from it (either to the left or the right), and none of these interceding digits are *d*. There is no smaller prime for which this is true. [Rivera]

733929337

The largest truncatable palindromic prime whose simultaneous deletion of one digit from each end always leaves a palindromic prime. [Gupta]

787080787

The smallest of only two existing nine-digit palindromic primes whose cube is a concatenation of three nine-digit primes:

$$\underbrace{487593529}\underbrace{299895709}\underbrace{745003403}.$$

Can you find the other? [De Geest]

799999999

A prime number formed by writing the number 7 followed by eight 9's. [Pinter and Karoly]

818752171
The smallest prime p such that the five numbers

$$\begin{aligned}&2^1+3^1+5^1+7^1+11^1+\ldots+p^1,\\&2^2+3^2+5^2+7^2+11^2+\ldots+p^2,\\&2^3+3^3+5^3+7^3+11^3+\ldots+p^3,\\&2^4+3^4+5^4+7^4+11^4+\ldots+p^4,\\&2^5+3^5+5^5+7^5+11^5+\ldots+p^5,\end{aligned}$$

are all primes. [Crespi de Valldaura]

933101339
The largest prime in "Luhn's Pyramid." All the palindromes below the top zero are primes.

0
101
31013
3310133
933101339
Luhn's Pyramid

964989469
The smallest palindromic prime that contains each of the composite digits only. [Beedassy]

968666869
The smallest palindromic prime with embedded beast number whose digits contain circles, i.e., using only the digits 0, 6, 8, 9. [De Geest]

987653201
The largest emirp with distinct digits. [Dale]

987654103
The largest prime with distinct digits. [McCranie]

1000004329
The smallest ten-digit prime that produces four other primes by changing only its first digit. [Opao]

1024383257
You cannot insert a digit in 1024383257 to form another prime. [Blanchette]

1057438801
Sierpiński (1956) conjectured that the Egyptian fraction $\frac{5}{n}=\frac{1}{a}+\frac{1}{b}+\frac{1}{c}$ could be solved (Guy, 1994). Bonnie M. Stewart confirmed this for all $n \leq 1057438801$.

1480028201	1480028129	1480028183
1480028153	1480028171	1480028189
1480028159	1480028213	1480028141

Figure 57. A Magic Square of Consecutive Primes

1111211111

The smallest triangle-congruent prime.

```
   1
  1 1
 1 2 1
1 1 1 1
```

1113443017

The first in a sequence of two dozen consecutive primes of the form $4n+1$. [Rivera]

1123465789

The smallest zeroless **pandigital prime**. [Weisstein]

The smallest 2-primeval prime, computed by Mike Keith in July 1998. All of the two-digit primes are embedded in the permutations of its digits. [Capelle]

1234567891

A zeroless consecutive-digit prime in ascending order. [Madachy]

1335557777

A prime in which the first, second, third, and fourth odd integers are repeated 1, 2, 3, and 4 times. [McGrath]

1477271183

The start of the smallest sequence of eleven consecutive emirps. [McCranie]

1480028171

The central prime of Harry Nelson's 3-by-3 magic square whose nine entries are consecutive primes. He won the $100 prize offered by Martin Gardner for finding it (Figure 57).

1500000001

J. van de Lune, H.J.J. te Riele, and D. T. Winter computed the first 1,500,000,001 complex zeros of the Riemann zeta function in 1986. They were all on the critical line just as the Riemann hypothesis suggests. [Beedassy]

1613902553

The start of eighteen consecutive primes with symmetrical gaps around the center. Figure 58 is a graphic display of the gaps between these primes. Note the gap of one between the twin primes at the center. [McCranie]

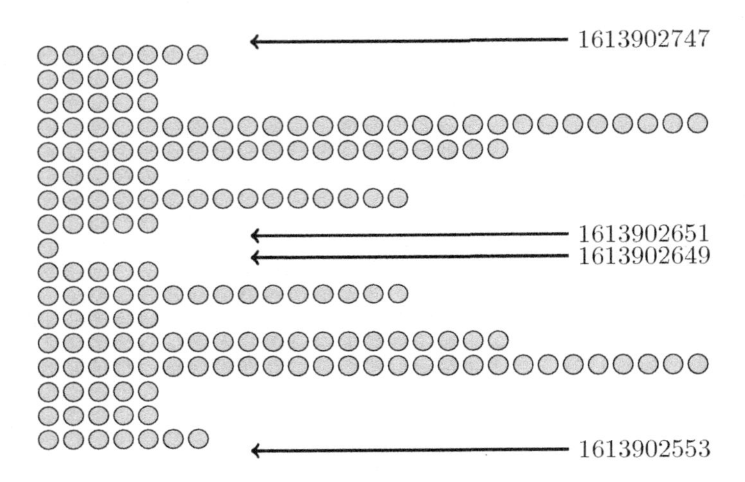

Figure 58. Symmetrical Prime Gaps Beginning at 1613902553

1617924853

The smallest zeroless pandigital prime whose reversal is a square. [Gupta]

1716336911

Delete any digit and it will remain prime. [Blanchette]

1719898171

1719898171 and $171^{98} + 98^{171}$ are primes. [Kulsha]

1882341361

The least prime whose reversal is both square and triangular. [Earls]

1966640443

The only known prime whose index (it is the 96664044th prime) is a substring (between its first and last digits). Note that the beast number is lurking inside. [Rivera]

Table 14. Smallest Prime with Given Multiplicative Persistence

persistence	prime	persistence	prime
0	2	6	8867
1	11	7	186889
2	29	8	2678789
3	47	9	26899889
4	277	10	3778888999
5	769	11	277777788888989

2106945901

If 210 is squared, and 45901 is squared, then the sum of these numbers is a prime with the digits 210 on the left and 45901 on the right. No other example has been found. [Haga]

2236133941

The first arithmetic progression of sixteen primes was found in 1969 by S. C. Root: $2236133941 + 223092870k$, where $k = 0$ to 15.

2999999929

A ten-digit prime such that a repdigit number is sandwiched between the tenth prime (i.e., 29). [Gupta]

3323333323

The largest **deletable prime** that is composed of only two digits. [Rupinski]

3592007533

The sum of the cubes of all three-digit palindromic primes. [Patterson]

3708797237

The start of an arithmetic progression of thirteen primes with a common difference equal to the product of the first thirteen primes. [McCranie]

3738668363

The first of two consecutive primes that are both emirps. [McCranie]

3778888999
The smallest prime with a multiplicative persistence of 10. (Persistence is defined in the first curio for 199.) See Table 14. [Gupta]

4076863487
The first of the smallest twin prime pair that adds to a fifth power (96^5). [Rivera and Trotter]

4332221111
One 4, two 3's, three 2's, four 1's is prime. [Kulsha]

4916253649
The smallest prime formed from the concatenation of six consecutive squares. [Gupta]

5838052333
$58^n + 38^n + 05^n + 23^n + 33^n$ is prime for $n = 0$ to 10. [Blanchette]

6607882123
The smallest member of the first Ormiston quadruple. [Andersen]

6771737983
A prime formed from five concatenated consecutive primes in ascending order. Another prime is formed if the same five primes are arranged in descending order, i.e., 8379737167. [Blanchette]

7427466391
What is the first 10-digit prime number occurring in consecutive digits of e? This question was posed on banners at a Cambridge, Massachusetts, subway stop and a billboard in California's Silicon Valley in July 2004 (Figure 59), as a cryptic pitch by *Google* to lure fresh talent.

$$\left\{ \begin{array}{c} \text{first 10-digit prime found} \\ \text{in consecutive digits of } e \end{array} \right\}\text{.com}$$

Figure 59. A Prime Billboard *Google* Ad (2004)

8018018851

The first prime that results from applying the rules of writing numbers in the American system of large number terminology as set out in Conway and Guy's *The Book of Numbers* (Springer-Verlag, 1996, p. 15): "eight billion eighteen million eighteen thousand eight hundred and fifty-one." Neil Copeland has suggested that 8,000,000,081 comes earlier, based on the spelling "eight billion and eighty-one."

8565705523

The (prime) number of digits in 1000000000!

9000000001

9000000001, 9900000001, 9990000001, and 9999000001 are each prime; but unfortunately, 9999900001 is composite.

9387802769

The start of the smallest sequence of twelve consecutive emirps. [McCranie]

9876543211

Clearly 9876543210 cannot be prime, so why not add 1? [Avrutin]

10000000019

The sum of the digits of the smallest eleven-digit prime is eleven. [Patterson]

10123457689

The smallest pandigital prime. [Weisstein]

10234465789

The smallest pandigital prime happy number (page 14). [Gupta]

10234786589

The lesser prime in the smallest pandigital twin prime pair. [Vrba]

10237573201

The smallest palindromic prime containing all of the non-composite digits. [Gupta]

10405071317

$1^1 + 2^2 + 3^3 + 4^4 + 5^5 + 6^6 + 7^7 + 8^8 + 9^9 + 10^{10}$. [Colucci]

10847395267

The smallest pandigital prime whose reversal is a square. [Gupta]

10985674123

It is conjectured that you can always arrange the numbers from 1 to $2n$ in a circle so that every pair of adjacent numbers adds to a prime. Antonio Filz called this circular permutation a Prime Circle. This prime gives one solution when $n = 5$.

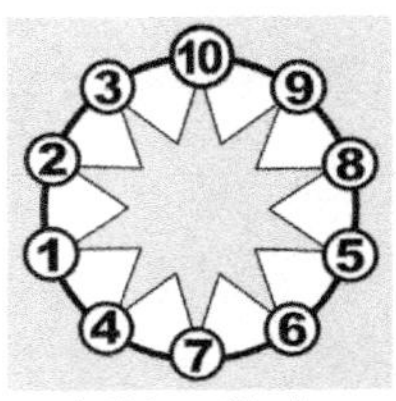

A Prime Circle

11172427111

The smallest palindromic prime such that the square of the sum of its digits equals the product of its digits. [Rivera and De Geest]

11373373373

The smallest natural number whose name is spelled with one hundred letters (eleven billion three hundred seventy-three million three hundred seventy-three thousand three hundred seventy-three). Can you find another example? [Brahinsky]

11853735811

Giovanni Resta discovered that 11853735811 is the 535252535th prime number. Is there a larger example where a prime and its index are each palindromic?

14141414141

The smallest smoothly undulating palindromic prime of the form $(14)_n1$. [Gallardo]

15984784979

Permute any two consecutive digits and you still have a prime number. [Blanchette]

18316337111

The phone number of "Prime Connections" (a company offering guided tours) in Castroville, CA, is 1(831)633-7111.

19896463921

The smallest prime starting a run of eight consecutive numbers of which the nth has exactly n prime factors. [Andersen]

19899999997

The largest possible prime phone number within the United States. Is it yours? [Rupinski]

19972667609

Describing 19972667609 and repeating the process with each new term produces six more primes, i.e., one 1, two 9's, one 7, one 2, two 6's, one 7, one 6, one 0, one 9, generates 1129171226171161019, etc. This record-breaking sequence was discovered by Walter Schneider.

34567876543

The largest of four palindromic primes composed of distinct consecutive digits listed first in ascending order and then in descending order. Omar E. Pol calls these the Giza primes (after their similarity to the pyramids at Giza, Egypt).

```
        8
       7·7
      6···6
     5·····5
    4·······4
   3·········3
```

37550402023

The sum of all primes less than one million is prime. [Honaker]

38358837677

Ramanujan noted that the inequality below holds if x is large enough. This is the largest known prime for which it fails.

$$\pi^2(x) < \frac{e \cdot x}{\log x}\, \pi\left(\frac{x}{e}\right)$$

[Berndt]

39713433671

The smallest prime in a "5TP39 twin prime cluster" (five twin primes in a group of thirty-nine consecutive numbers—the most that could be packed in an interval of this length). The name 5TP39 for these clusters (also called "quintuplet twin primes" by J. K. Andersen) was suggested by public librarian Roger Hargrave.

40144044691

The smallest prime that becomes composite if any digit is removed, changed, or inserted anywhere. [Andersen]

44560482149

The fifth odd prime Pell number is also a double Pell number of the first odd prime Pell number. The first odd prime Pell number is 5, $P(5) = 29$, and $P(P(5)) = P(29) = 44560482149$.

50006393431

The least example of a prime number in base ten that remains prime when expressed in bases two through nine (pretending they are in base ten). For example, in base seven this number is $(3420130221331)_7$, and in base ten 3420130221331 is a prime. [Brennen]

60335249959

The smallest prime that is the average of the ten surrounding primes (five on each side). [McCranie]

65639153579

$65639153579 \cdot 2^n + 2^n - 1$ is prime for $n = 0$ to 9. [Luhn]

98796959879

The most magical (largest) prime of the form *ABRACADABRA*. [Colucci]

99999199999

The largest palindromic prime less than a googol that has palindromic prime length. [Patterson]

100123456789

The smallest prime to contain the digit sequence 0123456789. [Gupta]

107928278317

The start of an eighteen-term arithmetic progression of primes with a common difference of 9922782870. [Pritchard]

111696081881

The smallest bemirp for which all four associated primes are the smaller members of twin prime pairs. [Cami]

125411328001

$1! \cdot 2! \cdot 3! \cdot 4! \cdot 5! \cdot 6! \cdot 7! + 1$. [Dobb]

129866728583

Truncate the rightmost digit and divide the result by two. Repeat using the number so obtained until left with a single digit. This is the largest prime such that repeating this procedure yields a prime at every step. [Rupinski]

131707310437
$1!^2 + 3!^2 + 5!^2 + 7!^2 + 9!^2$. [Post]

137438953481
Bertrand's postulate implies that there are infinitely many numbers α for which the following are all prime.

$$\lfloor 2^{\alpha} \rfloor, \lfloor 2^{2^{\alpha}} \rfloor, \lfloor 2^{2^{2^{\alpha}}} \rfloor, \lfloor 2^{2^{2^{2^{\alpha}}}} \rfloor, \lfloor 2^{2^{2^{2^{2^{\alpha}}}}} \rfloor, \ldots$$

The smallest of these is $\alpha = 1.2516475977904630175944320536 23...$ and generates the Bertrand primes: 2, 5, 37, 137438953481. The next Bertrand prime has 41,373,247,571 digits.

142112242123
Replacing each digit d with d copies of the digit d produces another prime throughout four transformations. [Jobling]

162536496481
A prime containing all double-digit squares in order. [Somer]

171727482881
The largest known elite prime. A prime number p is "elite" if only a finite number of Fermat numbers are squares modulo p (i.e., are quadratic residues of p). [Post]

198765432101
The smallest prime to contain the digit sequence 9876543210. [Gupta]

239651440411
The start of twenty-three consecutive full period primes (primes p for which $\frac{1}{p}$ has period length $p - 1$). [Andersen]

248857367251
The 9876543210th prime contains all of the digits except 9 and 0. [Necula]

274860381259
The smallest pandigital prime whose reversal is a cube. [Gupta]

275311670611
A difference of powers $(11^{11} - 10^{10})$. [Gupta]

308457624821
The number of thirteen-digit primes. [Dobb]

344980016453
The ASCII code for "PRiME" is the hexadecimal number 50 52 69 4D 45, which is the prime 344980016453 in base ten. (So PRiME is prime?) [Hartley]

351725765537
The concatenation of the five known Fermat primes 3, 5, 17, 257, and 65537. Concatenate them backwards, 655372571753, and it remains prime. [Rivera]

576529484441
The smallest prime formed from the reverse concatenation of four consecutive squares. [Gupta]

608888888809
The smallest strobogrammatic prime containing eight 8's. [Krussow]

608981813029
Prime numbers of the form $3n + 1$ are more numerous than those of the form $3n + 2$ at 608981813029. This is the very first time that this prime race favors $3n + 1$. [Bays and Hudson]

619737131179
The largest number such that every pair of consecutive digits is a different prime. (The answer to a question in *Eureka* 40, June 1979.)

689101181569
The start of another prime race sprint: twenty-nine consecutive primes of the form $4n + 1$. [Brennen]

902659997773
The smallest prime whose reciprocal has period length 666. [McCranie]

1099511628401
The largest number ever to be shown prime by Wilson's theorem (page 96) is 1099511628401. Because the work in finding 1099511628400! (even modulo 1099511628401) is horrendous, this record may stand for a long, long time! [Rupinski]

1111118111111

The smallest star-congruent prime of order 2. [Hartley]

```
      1
1   1   1   1
  1   8   1
1   1   1   1
      1
```

1131313515313

Each digit of this prime, read left-to-right, equals the number of composites between the first successive odd primes.

1226280710981

$1! - 2! + 3! - 4! + 5! - 6! + 7! - 8! + 9! - 10! + 11! - 12! + 13! - 14! + 15!$. [Warriner]

1618033308161

A palindromic prime obtained by reflecting the decimal expansion of the first seven digits of the golden ratio: $\phi = \frac{1+\sqrt{5}}{2} = 1.618033...$. [Necula]

1666666666661

The smallest prime containing exactly eleven 6's is palindromic. [De Geest]

1835211125381

The smallest palindromic prime that is also a Riesel number. [Rivera]

1888081808881

"This palindromic prime number reads the same upside down or when viewed in a mirror." From the top of the webpage *Patterns in Primes* by Harvey D. Heinz of Canada.

2748779069441

One of the largest known prime numbers at the end of the 19th century. It was found to be a factor of F_{36} by Seelhoff in 1886. [Dobb]

3059220303001

The smallest prime starting a run of nine consecutive integers for which the nth term has exactly n prime factors. [Andersen]

3111111111113

An easy-to-remember **depression prime** containing 13 digits. [Beisel]

3531577135439

The sum of primes less than or equal to 3531577135439 is a square. [Resta]

3763863863761

3763863863761 and $376^{386} \cdot 386^{376} + 1$ are both primes. [Kulsha]

6746328388801

Ramanujan listed all of the "highly composite numbers" less than 6746328388801, except one (293318625600). [Honaker]

6987191424553

The start of a run of sixteen consecutive primes, each ending with the same digit. [Resta]

7177111117717

The smallest palindromic prime such that the cube of the sum of its digits equals the product of its digits. [De Geest]

7625597485003

$2^{2^2} + 3^{3^3}$ is prime. [Kulsha]

9203225223029

The first multidigit palindromic prime equal to the sum of squares of three consecutive positive integers: $1751496^2 + 1751497^2 + 1751498^2$. [De Geest]

10102323454577

The smallest 14-digit prime with Shakespearean sonnet rhyme scheme (*ababcdcdefefgg*). [McCranie]

11108452651921

The Bernoulli triangle (see Table 15) is just like Pascal's triangle (see Table 16) except you start with powers of two $(2^0, 2^1, \ldots)$ down the right-hand side. Its second column $(2, 3, 4, \ldots)$ contains all of the primes, yet the smallest prime in any even-numbered column, other than column two, is 11108452651921 (14th column, 63rd row). [Edwards]

11410337850553

In 1993, Pritchard, Moran, and Thyssen found the first set of 22 primes in arithmetic progression; it begins with this prime and has a common difference of 4609098694200.

1								
1	2							
1	3	4						
1	4	7	8					
1	5	11	15	16				
1	6	16	26	31	32			
1	7	22	42	57	63	64		
1	8	29	64	99	120	127	128	
1	9	37	93	163	219	247	255	256

Table 15. Bernoulli Triangle

13301522971817

$1!^2 + 2!^2 + 3!^2 + 4!^2 + 5!^2 + 6!^2 + 7!^2 + 8!^2 + 9!^2 + 10!^2$. [Post]

18285670562881

The only known emirp to be formed by concatenating a row of Pascal's triangle (see the last row of Table 16). [Rivera]

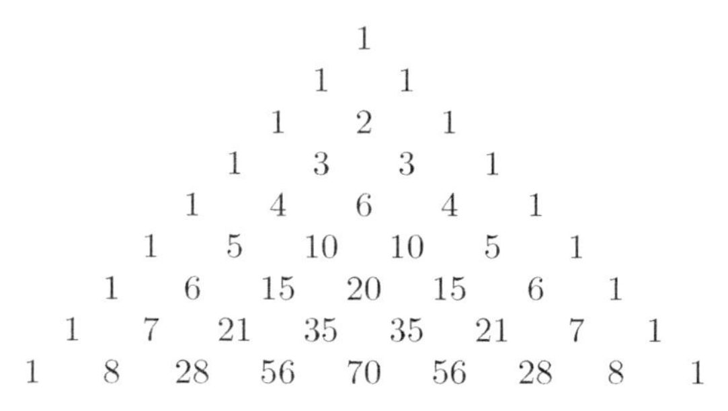

1								
1	1							
1	2	1						
1	3	3	1					
1	4	6	4	1				
1	5	10	10	5	1			
1	6	15	20	15	6	1		
1	7	21	35	35	21	7	1	
1	8	28	56	70	56	28	8	1

Table 16. Pascal's Triangle

23513892331597

A prime formed by concatenating the first seven primes of the Fibonacci sequence. [Rupinski]

28116440335967

The smallest multidigit prime narcissistic number, i.e., a number such that the sum of each digit raised to a power equal to the length of the number equals the number. Numbers with this property are sometimes known as Armstrong numbers or Pluperfect Digital Invariants (PPDI's).

29313741434753
The smallest prime formed from the concatenation of seven consecutive primes. [Gupta]

34445333343437
If the words in the KJV Bible's verses are replaced by their lengths, the first prime verse is Genesis 2:25 "And they were both naked, the man and his wife, and were not ashamed."

60111111111109
The smallest strobogrammatic prime containing all 1's between 60 and 09. [Gundrum]

74596893730427
The largest known prime repfigit number.

78456580281239
The smallest prime containing all ten digits that remains prime when all occurrences of any digit *d* are deleted for *d* from 0 through 9. [Resta]

81744303091421
The mean prime gap up to 81744303091421 is exactly 30. (See Table 11 on page 184.) [Carmody]

86812553978993
$98^7 + 654321$. [Hartley]

88929267773197
The largest left/right-truncatable prime. A left/right-truncatable prime is alternately truncatable, starting from the left. [Opao]

123467898764321
The largest palindromic mountain prime (page 69). [De Geest]

123571113171923
The concatenation of first ten non-composite numbers. [Gupta]

131175323571113
A prime number obtained by concatenating the consecutive primes thirteen down to two, and up to thirteen again. [Schlesinger]

134567897654321
The largest mountain prime (emirp). Its sum of digits is an emirp too! [Loungrides and Deel]

151515151515151
An easy-to-remember 15-digit smoothly undulating palindromic prime. [Honaker]

170669145704411
The start of the first occurrence of nine consecutive twin prime pairs. [DeVries]

303160419086407
The smallest prime that can be expressed as the sum of three distinct primes to their own powers $(7^7 + 11^{11} + 13^{13})$. [Patterson]

311111111111113
The third prime of the form $3(1)_k3$ has 13 ones. [Beisel]

320255973501901
The hexadecimal number `123456789ABCD` is prime. [Kulsha]

347811194367163
There are nine imaginary quadratic number fields with class number one and these have discriminants $-3, -4, -7, -8, -11, -19, -43, -67$, and -163. [Rupinski]

355693655479801
$(0 + 1)! + (2 + 3)! + (4 + 5)! + (6 + 7)! + (8 + 9)!$. [Post]

484511389338941
$n^2 + n + 484511389338941$ is prime for $n = 0$ to 13. [Luiroto]

535006138814359
In 1644, Mersenne wrote that $2^{257} - 1$ was prime. It took 283 years for Lehmer to prove him wrong; later, Penk and Baillie found its three prime factors. This is the smallest of those three.

870530414842019
The sum of the first ten million primes. [Vrba]

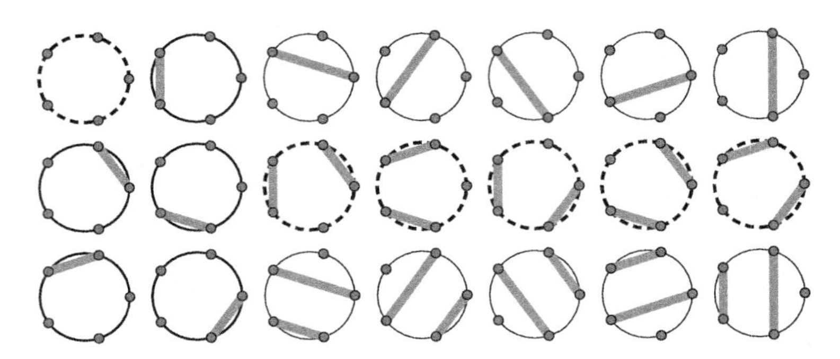

Figure 60. There are 21 Ways to Draw Nonintersecting Chords Between 5 Points on a Circle

953467954114363
The largest known prime Motzkin number. It is the number of different ways to draw nonintersecting chords on a circle between three dozen points (Figure 60). [Weisstein]

979853562951413
A prime whose reversal begins the decimal expansion of π. [Gardner]

999998727899999
The largest fifteen-digit palindromic prime that is the sum of three consecutive primes. [McCranie]

1011001110001111
The only known prime formed by the successive concatenation of an increasing number of repeated 1's and 0's. [Vrba]

1680588011350901
The greatest prime factor of any one hundred-digit repdigit.

1693182318746371
The smallest number that is followed by at least one thousand consecutive composites. It is the first prime gap of more than 1000 composites (actually 1131) and was discovered by a Swedish nuclear physicist (Dr. Bertil Nyman).

2744337540964913
The smallest prime formed from the concatenation of four consecutive cubes: 2744, 3375, 4096, and 4913. [Gupta]

3391382115599173
Write these digits as a 4-by-4 array, then each row, column, and the main diagonal form a 4-digit prime in both directions. [Blanchette]

3931520917431241
The smallest prime starting a run of ten consecutive integers for which the nth term has exactly n prime factors. [Andersen]

6171054912832631
$6171054912832631 + 366384 \cdot 23\# \cdot n$ is prime for n from 0 to 24. This is the first known arithmetic progression of primes of length 25 (see Table 17). It was found by Raanan Chermoni and Jaroslaw Wroblewski on May 17, 2008. [Jarek]

6664666366626661
There are four Horsemen of the Apocalypse mentioned in chapter 6 of the Book of Revelation. This prime presents a "beastly" countdown from 4. [Patterson]

8008808808808813
Alphabetically the first prime in German (acht Billiarden, acht Billionen, achthundertacht Milliarden, achthundertacht Millionen, achthundertachttausendachthundertdreizehn). [Raab]

8690333381690951
Marxen and Buntrock proved in 1997 that the maximal number of steps that a six-state Turing machine can make on an initially blank tape before eventually halting is at least 8690333381690951.

Alan Turing Stamp

9999999900000001
91 is composite and 9901 is prime. 999001 is composite and 99990001 is prime. 9999900001 is composite and 999999000001 is prime. 99999990000001 is composite and 9999999900000001 is prime. 999999999000000001 is composite and, well, so is the next term—but it was nice while it lasted. [Beiler]

10269797835402631
Delete (one at a time) the zeros of this pandigital prime in any order; continue in this manner for the threes, sixes, and nines, to form a dozen different primes. [Blanchette]

Table 17. Smallest k-Term Arithmetic Progressions of Primes

k	Primes	Year	Who
3	$3 + 2n$		
4	$5 + 6n$		
5	$5 + 6n$		
6	$7 + 30n$	1909	[1]
7	$7 + 150n$	1909	[1]
8	$199 + 210n$	1910	[2]
9	$199 + 210n$	1910	[2]
10	$199 + 210n$	1910	[2]
11	$110437 + 13860n$	1967	[3]
12	$110437 + 13860n$	1967	[3]
13	$4943 + 60060n$	1963	[4]
14	$31385539 + 420420n$	1983	[5]
15	$115453391 + 4144140n$	1983	[5]
16	$53297929 + 9699690n$	1976	[6]
17	$3430751869 + 87297210n$	1977	[6]
18	$4808316343 + 717777060n$	1983	[5]
19	$8297644387 + 4180566390n$	1984	[5]
20	$214861583621 + 18846497670n$	1987	[7]
21	$5749146449311 + 26004868890n$	1992	[5]
22	$11410337850553 + 4609098694200n$	1993	[8]
23	$403185216600637 + 2124513401010n$	2006	[9]
24	$515486946529943 + 30526020494970n$	2008	[10]
25	$6171054912832631 + 81737658082080n$	2008	[10]

Notes: Each of these forms yield primes for n from 0 to $k-1$. They are proven to be the arithmetic progressions of primes with least final term for $n \leq 21$, and are the least known for the four others.

Discoverers: [1] G. Lemaire, [2] Edward B. Escott, [3] Edgar Karst, [4] V. N. Seredinskij, [5] Paul Pritchard, [6] Sol Weintraub, [7] Jeff Young & James Fry, [8] Paul Pritchard and others, [9] Markus Frind, and [10] Raanan Chermoni & Jaroslaw Wroblewski.

Granville estimates that the largest prime in the smallest k-term arithmetic progression is about $(e^{1-\gamma}k/2)^{k/2}$, where γ is the Euler-Mascheroni constant.

See Andersen's *Primes in Arithmetic Progression Records* listed in the Prime Sites (page 268).

113335555577777777

Eric Sorensen found this stuttering prime on his Commodore 64. (Remember the home computers of 1982?)

12348516587712457

A prime formed on a Sudoku puzzle from what may be a minimum number of hints (Figure 61).

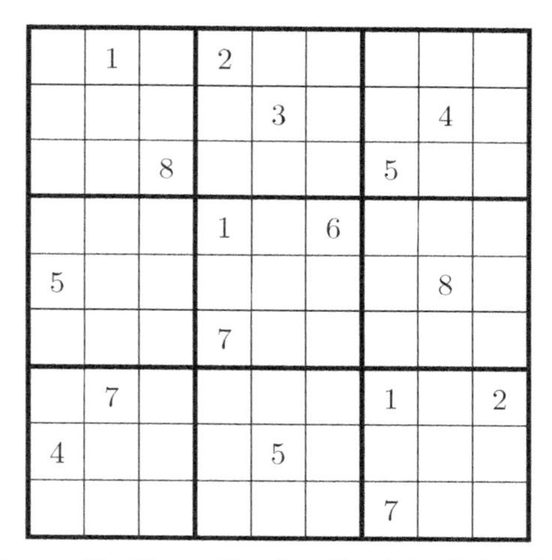

	1		2					
				3			4	
		8				5		
			1		6			
5							8	
			7					
	7					1		2
4				5				
						7		

Figure 61. From Gordon Royle's Collection

13311464115101051

The smallest emirp formed by concatenating consecutive rows of Pascal's triangle (see Table 16 on page 215). [Firoozbakht]

17000000000000071

Could there be an easier-to-remember 17-digit palindromic prime? [Gregor]

22439962446379651

The first prime p for which the next prime is exactly 1000 larger.

59604644783353249

The smallest prime of the form $5^n + n^5$. [Patterson]

66555444443333333

The number of times each digit of this prime is repeated corresponds to the first four primes. [Sorensen]

71828182828182817
A palindromic prime formed from the reflected decimal expansion of '$e - 2$'. [Kulsha]

97654321012345679
The largest palindromic prime with strictly decreasing digits up to the middle, and then strictly increasing. [De Geest]

98823199699242779
The smallest odd prime that can be represented as the sum of a Fibonacci number and its reversal, i.e., 37889062373143906 plus 60934137326098873. [Gupta]

261798184036870849
A devilish prime containing 6 + 6 + 6 digits that can be expressed as $666^6 + 666^6 + 666^6 + 1$. [Patterson]

496481100121144169
A prime formed by concatenating 7 consecutive squares, starting with 7^2. [Mendes]

909090909090909091
$\frac{10^{19}+1}{11}$. (Like the number itself, this contains only the digits 0, 1, and 9.) [Pickover]

1018264464644628101
A palindromic prime of the form $k^2 + 1$. [De Geest]

1023456987896543201
The smallest pandigital palindromic prime. It was proven prime by Harry L. Nelson in 1980.

1111001101011001111
The smallest centered hexagonal-congruent prime of order 3. [Hartley]

```
    1   1   1
  1   0   0   1
1   0   1   0   1
  1   0   0   1
    1   1   1
```

1147797409030815779
The largest prime with distinct digits in hexadecimal. It is equivalent to FEDCBA987654023. [Kulsha]

1234567654321234567
A curious N-shaped prime. [Necula]

12345678949876543 21

An easy-to-remember palindromic prime whose central digit (the first composite) is the only digit that makes this number a prime. [Mendes]

1294584340434854921

The smallest palindromic prime that remains prime if the middle digit is replaced by each of six other digits (in this case 2, 3, 5, 6, 8, or 9).

1341351362137138139

The number $\underbrace{134}\underbrace{135}\underbrace{136}\,\mathbf{2}\,\underbrace{137}\underbrace{138}\underbrace{139}$ is the first of three primes in arithmetic progression. To find the other two, just replace the middle digit 2 with 5, and then with 8. [Mendes]

1524157875019052101

$1234567890^2 + 1$ is prime. [Kik]

1712329866165608783

The smallest prime of the form $n^2 + n + 1712329866165608771$. This quadratic form is expected to produce a plethora (high-density) of prime numbers.

1799999999999999999

An easy-to-remember prime containing 17 9's.

2305843009213693951

Wonder Book of Strange Facts (Ripley's Believe It or Not, Inc., 1957, p. 100) states that this (the 9th Mersenne prime) is the number of ways to make change for a five-dollar bill using pennies, nickels, dimes, quarters, half-dollar, and dollar coins. The correct answer is 98411, making this one of the largest prime errors ever found. [Rupinski]

3203000719597029781

The start of a record-breaking Cunningham chain (2nd kind). It has length 16 and was found by Tony Forbes on December 5, 1997.

3331113965338635107

The first prime reached if you concatenate the prime factors of 8 and repeat the process. Primes for any such integer have been called home primes. For example,

$$8 \mapsto 2 \cdot 2 \cdot 2 \mapsto 2 \cdot 3 \cdot 37 \mapsto 3 \cdot 19 \cdot 41 \mapsto 3 \cdot 3 \cdot 3 \cdot 7 \cdot 13 \cdot 13 \mapsto 3 \cdot 11123771 \mapsto 7 \cdot 149 \cdot 317 \cdot 941 \mapsto 229 \cdot 31219729 \mapsto 11 \cdot 2084656339 \mapsto 3 \cdot 347 \cdot 911 \cdot 118189 \mapsto 11 \cdot 613 \cdot 496501723 \mapsto 97 \cdot 130517 \cdot 917327 \mapsto 53 \cdot 1832651281459 \mapsto 3 \cdot 3 \cdot 3 \cdot 11 \cdot 139 \cdot 653 \cdot 3863 \cdot 5107.$$

6082394749206781697
$1!^{11} + 2!^{10} + 3!^{9} + 4!^{8} + 5!^{7} + 6!^{6} + 7!^{5} + 8!^{4} + 9!^{3} + 10!^{2} + 11!^{1}$. [Post]

12345678901234567891
A prime whose digits are in "ascending order." [Madachy]

22335577111113131717
The smallest prime obtained by repeating the first n consecutive primes. [Gallardo]

43252003274489855999
The number of different unsolved configurations that can be reached on the original Rubik's Cube. [Honaker]

44211790234832169331
The 10^{18}th prime number (see Table 18). [Dobb]

51091297865364919801
$(0+1)! + (2+3)! + (4+5)! + (6+7)! + (8+9)! + (10+11)!$. [Post]

71322723161814151019
The largest self-descriptive prime without repetition. [Rivera]

89726156799336363541
The largest left-truncatable prime ending with the digit 1, if we allow 1 to be considered a prime. [Weichsel]

99999999999999999989
The largest 20-digit prime number.

100112233445566877989
The smallest prime that contains each of the digits a prime number of times. [Gallardo]

Table 18. The 10^kth Prime

n	nth prime
1	2
10	29
100	541
1,000	7,919
10,000	104,729
100,000	1,299,709
1,000,000	15,485,863
10,000,000	179,424,673
100,000,000	2,038,074,743
1,000,000,000	22,801,763,489
10,000,000,000	252,097,800,623
100,000,000,000	2,760,727,302,517
1,000,000,000,000	29,996,224,275,833
10,000,000,000,000	323,780,508,946,331
100,000,000,000,000	3,475,385,758,524,527
1,000,000,000,000,000	37,124,508,045,065,437
10,000,000,000,000,000	394,906,913,903,735,329
100,000,000,000,000,000	4,185,296,581,467,695,669
1,000,000,000,000,000,000	44,211,790,234,832,169,331
10,000,000,000,000,000,000	465,675,465,116,607,065,549

123456789878987654321

A palindromic prime that contains a triangular peak and crater. [Necula]

147573952589676412931

The largest known prime of the form $2^p + 3$, where p is prime.

157158159160161162163

The smallest prime formed by concatenating n consecutive increasing numbers starting with an odd prime (157) and ending with the next consecutive prime (163). [Honaker]

283453407513524913023

$37^2 + 31^3 + 29^5 + 23^7 + 19^{11} + 17^{13} + 13^{17} + 11^{19} + 7^{23} + 5^{29} + 3^{31} + 2^{37}$. [Silva]

455666777788888999999

One 4, two 5's, three 6's, four 7's, five 8's, six 9's is prime. [Mendes]

827125343133121974913

The concatenation of the cubes of the primes from 2 to 17 is prime. [Patterson]

12345678910109876543 21

A generalized Smarandache palindrome (GSP) is any integer of the form $a_1a_2a_3\ldots a_na_n\ldots a_3a_2a_1$ or $a_1a_2a_3\ldots a_{n-1}a_na_{n-1}\ldots a_3a_2a_1$, where all $a_1, a_2, a_3, \ldots, a_n$ are positive integers of various number of digits. 1234567891010987654321 is the largest known prime GSP, if $a_1, a_2, a_3, \ldots, a_n$ represent the sequence of natural numbers.

1235607889460606009419

The smallest prime that can be reduced to every one of the twenty-six minimal primes in base ten by removing digits. (See the entry for 66600049.) [Rupinski]

2030507011013017019023

A prime made up from the digits of the first nine primes interlaced with single zeros. [Sorensen]

4669201609102990671853

The smallest prime formed by the leading digits of the decimal expansion of the first Feigenbaum constant (4.66920...). (The two Feigenbaum constants arise in chaos theory.)

5021837752995317770489

$987^6 \cdot 5432 + 1$ is prime. [Kulsha]

5471619276639877320977

The final prime in the smallest arithmetic progression of 17 primes beginning with 17 (see Table 19). [Granville]

23333333333333333333333

An easy-to-remember 23-digit prime of the form $2(3)_k$. [Daugherty]

29998999999999999999999

The smallest prime with an additive persistence (defined in the first curio for 199) equal to the smallest composite number. Can you explain why this book doesn't list a prime with persistence one higher? [Gupta]

Table 19. Smallest p-term Arithmetic Progression of Primes Beginning With p

p	Arithmetic Progression	Last Term
2	$2 + n$	3
3	$3 + 2n$	7
5	$5 + 6n$	29
7	$7 + 150n$	907
11	$11 + 1536160080n$	15361600811
13	$13 + 9918821194590n$	119025854335093
17	$17 + 341976204789992332560n$	5471619276639877320977

35452590104031691935943

The largest prime narcissistic number (page 215). [De Geest]

46931635677864055013377

The first known factor of F_{31} was discovered by A. Kruppa on April 12, 2001, using a sieving program developed by T. Forbes. (The Fermat numbers grow so fast that our only hope of showing them not prime is by finding a factor.)

98765432101310123456789

The smallest palindromic prime containing 9876543210 on the left. [Seidov]

124829153113731171731393

An exterior truncatable prime. Take digits off at each end until you get down to two digits. All subsequent steps are primes. [Patterson]

357686312646216567629137

The largest left-truncatable prime.

396876511751700883754879

$101 \cdot 202 \cdot 303 \cdot 404 \cdot 505 \cdot 606 \cdot 707 \cdot 808 \cdot 909 - 1$ is prime. [Poo Sung]

1111118881188811888111111

The only strobogrammatic and tetradic square-congruent prime of order 5. [Ward]

```
1 1 1 1 1
1 8 8 8 1
1 8 8 8 1
1 8 8 8 1
1 1 1 1 1
```

1223334444555554444333221

A palindromic prime formed from the first five digits, where each digit d is repeated d times in ascending and descending order. [Silva]

1232123421234323453234543

Reading left-to-right, each digit gives you the number of letters required in the Roman numeral representation from I to XXV.

1313131313131313131313131

A prime formed by concatenating thirteen 13's, then deleting the last digit. [Avrutin]

1997392401978791042937991

Arrange this palindromic prime into a 5-by-5 array and each row, column, and main diagonal will form an emirp. [Blanchette]

1	9	9	7	3
9	2	4	0	1
9	7	8	7	9
1	0	4	2	9
3	7	9	9	1

2000000000000000000000003

The only prime formed by inserting p 0's between a two-digit prime p. [Silva]

2378741380049115007971348 1

The number of dissections of a convex 20-gon by nonintersecting diagonals into an odd number of regions. [Post]

4786774223206688004761107 9

In 1975, F. Cohen and J. L. Selfridge showed that this prime plus or minus a power of 2 can never be a prime or a power.

8056166352780240625732174 7

Ramanujan's $\tau(n)$ function is composite for $2 \leq n \leq 63000$. It is first prime at $n = 63001$. Ramanujan's tau function is defined by the following.

$$x \prod_{m=1}^{\infty} (1 - x^m)^{24} = \sum_{n=1}^{\infty} \tau(n) x^n$$

6189700196426901374495621 11

The smallest Mersenne prime containing all of the digits from 0 to 9. [Gupta]

742950290870000078092059247

The first prime in an arithmetic sequence of ten palindromic primes. It has a common difference of $10101 \cdot 10^{11}$, and was found by Harvey Dubner and his assistants. [Trotter]

77777777777771313131313131313

Thirteen 7's followed by seven 13's is prime. "Lucky 7" and "Unlucky 13" keeping each other in perfect balance. [De Geest]

8977777777777777777777777689

A prime factor of one googolplex + 10, i.e., $10^{10^{100}} + 10$. [Broadhurst]

916282611716282617116282619

A palindromic prime that can be extracted from the concatenation of four consecutive primes: 16282610**9**, **162826117**, **162826171**, and **16282619**9. [De Geest]

984417984040090040489714489

A palindromic prime of the form $k + (k + 1)^2$, where k is 31375435997608. [De Geest]

1234567891234567891234567891

One of the primes in the Smarandache deconstructive sequence of integers, which is constructed by sequentially repeating the digits (from 1 to 9) in the following way: 1, 23, 456, 7891, 23456, 789123, 4567891, 23456789, 123456789, 1234567891, ...

6660000000000000000000000007

The smallest non-palindromic beastly prime with a 7 at the right end. (The 2nd such prime occurs when this number's two dozen consecutive zeros are increased by one.)

9419919379719113739773313173

The largest prime such that any three consecutive digits is a distinct prime. [Opao]

1717171717171717171717171717171

The smallest prime of the form $1(71)_k$. [Adam]

8939662423123592347173339993799

The largest deletable prime for which a digit can always be deleted from one of the two ends, leaving a smaller prime after each deletion. [Andersen]

41117342095090841723228045851817

$1!^2 + 2!^2 + 3!^2 + 4!^2 + 5!^2 + 6!^2 + 7!^2 + 8!^2 + 9!^2 + 10!^2 + \ldots + 17!^2 + 18!^2$. [Post]

162259276829213363391578010288127

In 1914, R. E. Powers announced that he had found $M(107)$ to be prime and almost immediately E. Fauquembergue announced he also had discovered this fact. Note that 1914 was the start of WWI!

425260376469080434957375449438061

The smallest prime formed from the concatenation of the first n terms of "Madonna's sequence," which is generated by adding 1 (modulo 10) to each digit in the decimal expansion of π (if the digit is 9, it becomes a 0). [Dowdy]

1566921454155358203906337037634649

Changing each letter of the word "MATHEMATICALLY" to ASCII Code (in binary blocks of eight), and then converting it to decimal yields a prime. Curiously, "mathematically" gives an even larger prime. [Jobling]

4316720792370367095095683949638501

$29^{23} + 23^{19} + 19^{17} + 17^{13} + 13^{11} + 11^{7} + 7^{5} + 5^{3} + 3^{2}$. [Silva]

9551979199733313739311933719319133

The largest prime such that any four adjacent digits are distinct primes. It was found by Luke Pebody, who is better known for solving the necklace problem in number theory.

18133392183093337273339038129333181

Note that (beginning with 2) each successive number in Table 20 is the smallest palindromic prime to contain the one above it as its center. This prime is the 11th in what is surely an infinite pyramid.

68476562763327854359085599065855383

$2^{2+3\cdot5\cdot7} + 3^{3+2\cdot5\cdot7} + 5^{5+2\cdot3\cdot7} + 7^{7+2\cdot3\cdot5}$ is prime (as are the four exponents). [Rupinski]

9000000999999999994999999999990000009

This is the largest of the set of eighty-six minimal palindromic primes. Every palindromic prime can be reduced to one of these by removing zero or more symmetric pairs of digits. The list begins 2, 3, 5, 7, 11, 919, 94049, 94649, 94849, 94949, 96469, 98689, ...

2

727

37273

333727333

93337273339

309333727333903

1830933372733390381

92183093337273339038129

3921830933372733390381293

1333921830933372733390381293331

18133392183093337273339038129333181

11718133392183093337273339038129333181711

171171813339218309333727333903812933318171171

1517117181333921830933372733390381293331817117151

3461517117181333921830933372733390381293331817117151643

93346151711718133392183093337273339038129333181711715164339

339334615171171813339218309333727333903812933318171171516433933

180339334615171171813339218309333727333903812933318171171516433933081

120180339334615171171813339218309333727333903812933318171171516433933081021

194120180339334615171171813339218309333727333903812933318171171516433933081021491

126194120180339334615171171813339218309333727333903812933318171171516433933081021491621

336126194120180339334615171171813339218309333727333903812933318171171516433933081021491621633

331336126194120180339334615171171813339218309333727333903812933318171171516433933081021491621633133

Table 20. The Smallest Nested Palindromic Primes (see page 230)

31415926535897932384626433832795028841
A prime number embedded in the decimal expansion of π. [Baillie]

43143988327398957279342419750374600193
The smallest Leyland prime (i.e., of the form $x^y + y^x$, where x and y are natural numbers, $1 < x \leq y$) with x and y both composite numbers ($x = 15$, $y = 32$). [Beedassy]

10111213141516171819181716151413121101
Recipe for a palindromic prime: write down the palindrome 0123456789876543210 and separate each digit with a 1. Include a 1 on each end. Found by Bobby Jacobs (a math whiz from Virginia), who was almost 9-years-old at the time.

170141183460469231731687303715884105727
The Mersenne prime $2^{127} - 1$ that Édouard Lucas verified in 1876. He is said to have spent 19 years in checking it, by hand. This remains the largest prime number discovered without the aid of a computer. Lucas died under unusual circumstances. A banquet waiter dropped a plate and a broken piece flew up and cut him on the cheek. He passed away a few days later from a bacterial infection (erysipelas).

Lucas (1842–1891)

191918080818091909090909190818080819191
This palindromic Sophie Germain prime starts a Cunningham chain of three palindromic primes. [Dubner]

958619577835947143938319314151899378973
Omit the rightmost digit from this prime and the remaining number is the largest right-truncatable semiprime. [Gupta and Honaker]

1894741890219724182316273512181213141511
A left-truncatable prime of order two. By taking off two digits at a time, from the left, the number stays prime until you get down to two digits. [Patterson]

2616218222822143606864564493635469851817
$1!^2 + 2!^2 + 3!^2 + 4!^2 + 5!^2 + 6!^2 + \ldots + 20!^2 + 21!^2$. [Post]

1193331618151217133020331712151816133391 1

The largest palindromic prime in a palindromic prime pyramid of step size two, where each palindromic prime is the smallest palindromic prime that contains the palindromic prime in the row above (Figure 62). Note that "11" appears on the ends of row 11, reminding us that all palindromic numbers greater than 11, containing an even number of digits, are divisible by 11.

2
30203
133020331
1713302033171
12171330203317121
151217133020331712151
1815121713302033171215181
16181512171330203317121518161
331618151217133020331712151816133
9333161815121713302033171215181613339
11933316181512171330203317121518161333911

Figure 62. Palindromic Prime Pyramid

39916800362880036288040320504072012024621 1

The concatenation of the first dozen factorials starting with 0! in reverse order. [Hartley]

10000000000000000000000000000000000000000063

The smallest prime greater than an American tredecillion (or British septillion). It is also an emirp. [Patterson]

13763761774552805635707475936358698644919801

$(0+1)! + (2+3)! + (4+5)! + (6+7)! + (8+9)! + \ldots + (18+19)!$. [Post]

20988936657440586486151264256610222593863921

The Frenchman Aimé Ferrier (author of the book *Les Nombres Premiers*, 1947) used a mechanical desk calculator to find this prime in 1951, making it the largest prime found before electronic computers.

42000000000000000000000000000000000000000043

A prime number that begins with 42, ends with 43, and has a length of 44 digits. This is also the only known prime of the form $n \cdot 10^n + n + 1$. [Earls]

999999999888888887777777666666555554444223343

The largest prime p such that every decimal digit d appears exactly d times. [Rivera]

2084924134437040933465549100652627305664757 81

$1^1 + 2^2 + 3^3 + \ldots + 28^{28} + 29^{29} + 30^{30}$ is prime. [Brod]

393050634124102232869567034555427371542904833

The second **Cullen prime**: $141 \cdot 2^{141} + 1$. A Cullen prime of the form $p \cdot 2^p + 1$, where p is prime, has never been found.

568972471024107865287021434301977158534824481

The largest known prime of the form $\frac{p^p - 1}{p-1}$. [Rupinski]

802359150003121605557551380867519560344356971

The first in a quadruplet of primes p, $p+2$, $p+6$, and $p+8$. [Penk]

888888888888888888888888888888888888888888899

The smallest prime in the sequence 99, 899, 8899, 88899, 888899, etc.

1709172117231733174117471753175917771783178717 89

The ordered-concatenated prime years of the 18th century is prime. [Gallardo]

7777777999997796669779686977966697799999777 7777

A square-congruent prime that is the sum of three other square-congruent primes (Figure 63). [Hartley]

1 1 1 1 1 1 1		3 3 3 3 3 3 3		3 3 3 3 3 3 3		7 7 7 7 7 7 7
1 2 2 2 2 2 1		3 0 0 0 0 0 3		3 7 7 7 7 7 3		7 9 9 9 9 9 7
1 2 0 0 0 2 1		3 0 6 6 6 0 3		3 7 0 0 0 7 3		7 9 6 6 6 9 7
1 2 0 2 0 2 1	+	3 0 6 5 6 0 3	+	3 7 0 1 0 7 3	=	7 9 6 8 6 9 7
1 2 0 0 0 2 1		3 0 6 6 6 0 3		3 7 0 0 0 7 3		7 9 6 6 6 9 7
1 2 2 2 2 2 1		3 0 0 0 0 0 3		3 7 7 7 7 7 3		7 9 9 9 9 9 7
1 1 1 1 1 1 1		3 3 3 3 3 3 3		3 3 3 3 3 3 3		7 7 7 7 7 7 7

Figure 63. A Sum of Square-Congruent Primes

3748313799397919391139979319913931333179393719997 13

The largest prime such that any five adjacent digits form a distinct prime, if we do not allow leading zeros. [Andersen]

(1)
13
13**9**
1**27**39
12**81**739
1281739**243**
128173**729**9243
12817372**2187**99243
1281737**6561**2218799243
12**19683**8173765612218799243
1219683817376**59049**5612218799243
12196838**177147**17376590495612218799243
12196838**531441**17714717376590495612218799243
121968385314 41**1594323**17714717376590495612218799243
121968385314411594323177147**4782969**17376590495612218799243

Figure 64. The Tower of Power

3141592653589793238462643383346264832397985356295141 3

A palindromic prime formed from the reflected decimal expansion of π. [Honaker]

**71111111111111111111111111111111111111111111111111\
1111**

The largest known prime of the form $7(1)_k$.

**1219683853144115943231771474782969173765904956122187\
99243**

Beginning at one and successively inserting the powers of three we may form a pyramid of primes (Figure 64). How long can this continue? (Probably forever.) [Rupinski]

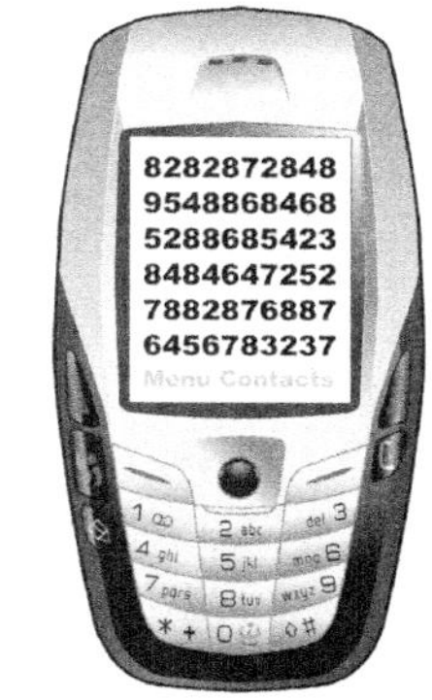

Cypher Prime

**82828728489548868468528868542384846 47\
2527882876887645678 3237**

The largest "Cypher prime" in the KJV Bible; these integers are named after the movie *Cypher* (2002), where the hero is secretly given a phone number via a Bible verse. If you enter the first letter of each word of 1 Kings 15:18 into a standard mobile phone, then the corresponding number will appear on the screen. [Bulmer]

**1432551529933067325229732842795357013757693555710038\
47375258761**
A prime also written as $\frac{23^{47}+47^{23}}{23+47}$. [Kulsha]

**2056880696651507552693711478196688131228419832047112\
81293004769**
The expression $69^8 + 8^{69}$ remains the same when turned upside down. [Kulsha]

**5213237242362344323423653133643623435234341436341336\
46239119311**
If the words in the KJV Bible's verses are replaced by their lengths, the largest prime verse is Daniel 5:11.

**1234567891234567891234567891234567891234567891234567\
891234567891234567**
The sequence 123456789 repeated seven times, followed by 1234567, is prime. [Goltz]

**2030507011013017019023029031037041043047053059061067\
071073079083089097**
A prime number containing the sequence of primes less than one hundred, each separated by a single zero. [Honaker]

**2998715143089348424671669750806424093608691761461607\
292446410696603059 7**

2# + 3# + 5# + 7# + 11# + ... + 173# + 179# − 1 is prime, as are the next two terms in this series of prime primorials. [Rupinski]

**5210644015679228794060694325390955853335898483908056\
4583521838510183725557352 21**

$180(2^{127} - 1)^2 + 1$. The first prime proven by computer to hold the title of largest known prime. It was discovered in 1951 by J.C.P. Miller and D. J. Wheeler, using a British computer called EDSAC (Electronic Delay Storage Automatic Calculator).

**1009969724697142476377866555879698403295093246891900\
418036034177589043417033488821590672297 19**
The prime that starts the longest known sequence of consecutive primes in arithmetic progression. The ten primes have a common

3 1 3 9 9 7 1 9 7 3
7 8 6 6 3 4 7 1 1 3
9 1 4 4 8 6 5 1 5 7
7 2 6 9 4 8 5 8 9 1
7 5 9 4 1 9 1 2 2 9
3 8 7 4 4 5 9 1 8 7
7 6 5 6 9 2 5 7 8 9
7 4 7 9 7 4 9 1 4 3
1 9 4 2 2 8 8 9 6 1
1 3 7 3 9 3 9 7 3 1

Figure 65. Reversible Prime or Primes?

difference of 210 and were found by Manfred Toplic on March 2, 1998. A search for an arithmetic progression of eleven consecutive primes will be difficult, where the minimum gap between these primes is 2310 instead of 210.

1414213562373095048801688724209698078569671875376948
0731766797379907324784621070388503875343276 41

A prime formed from the first 97 digits of the decimal expansion of $\sqrt{2}$. [Gupta]

3139971973786634711391448651577269485891759419122938
7445918776569257897479749143194228896113739397 31

Write this reversible prime in a 10-by-10 square. All rows, columns, and main diagonals are distinct reversible primes (Figure 65). This means the reverse 100-digit prime could have been submitted instead! [Andersen]

7777277227777723277...(121 digits)...7723277777227727777

The smallest star-congruent prime (see Figure 66) containing all four prime digits. [Hartley]

7994412097716110548 1...(154 digits)...4315422474838221101 9

The smaller of two primes that Roger Schlafly patented in 1994 for use with his algorithm for modular reduction (U.S. Patent 5373560).

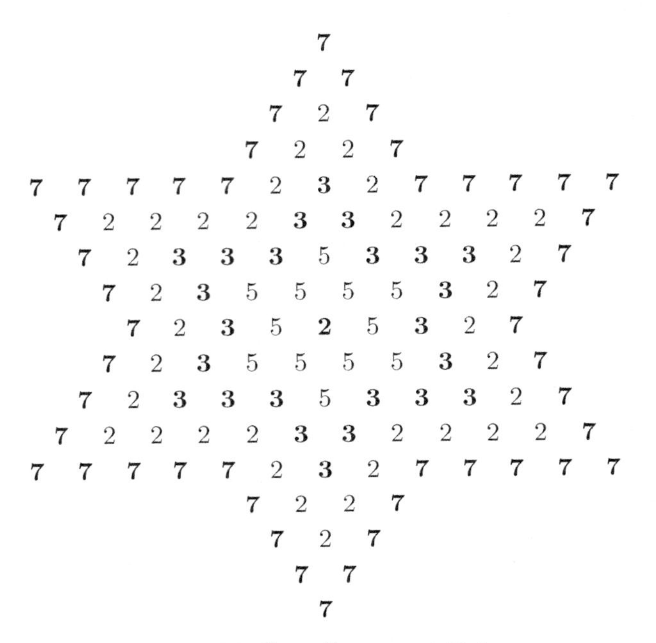

Figure 66. Star-Congruent Prime

8281807978777675747 3...(155 digits)...51413121110987654321
The only known prime in the Smarandache reverse sequence, i.e., $1, 21, 321, 4321, \ldots, 10987654321$, etc. Micha Fleuren searched up to the first 10000 terms and found no others. [Gregory]

1672375728362281764 3...(200 digits)...9999999999999999999
Frank Morgan's "Math Chat" column (1997) challenged readers to find the largest prime number by using four 4's and a finite number of mathematical symbols and operators in common use (see the four 4's puzzle on page 54). William Foster had the best submission: $((\sqrt{4}/.4)!! - .4)/.4$. [Brahinsky]

4856507896573978293...(1401 digits)...3476861420880529443
This was the first known **illegal prime**. What people often forget is that a program (any file actually) is a string of bits (binary digits), so every program is a number. Some of these are prime. Phil Carmody found this one in March 2001. When written in base 16

4

8565078965 7397829309 8418946942 8613770744 2087351357
9240196520 7366869851 3401047237 4469687974 3992611751
0973777701 0274475280 4905883138 4037549709 9879096539
5522701171 2157025974 6669932402 2683459661 9606034851
7424977358 4685188556 7457025712 5474999648 2194184655
7100841190 8625971694 7970799152 0048667099 7592359606
1320725973 7979936188 6063169144 7358830024 5336972781
8139147979 5551339994 9394882899 8469178361 0018259789
0103160196 1835034344 8956870538 4520853804 5842415654
8248893338 0474758711 2833959896 8522325446 0840897111
9771276941 2079586244 0547161321 0050064598 2017696177
1809478113 6220027234 4827224932 3259547234 6880029277
7649790614 8129840428 3457201463 4896854716 9082354737
8356619721 8622496943 1622716663 9390554302 4156473292
4855248991 2257394665 4862714048 2117138124 3882177176
0298412552 4464744505 5834628144 8833563190 2725319590
4392838737 6407391689 1257924055 0156208897 8716337599
9107887084 9081590975 4801928576 8451988596 3053238234
9055809203 2999603234 4711407760 1984716353 1161713078
5760848622 3637028357 0104961259 5681846785 9653331007
7017991614 6744725492 7283348691 6000647585 9174627812
1269007351 8309241530 1063028932 9566584366 2000800476
7789679843 8209079761 9859493646 3093805863 3672146969
5975027968 7712057249 9666698056 1453382074 1203159337
7030994915 2746918356 5937621022 2006812679 8273445760
9380203044 7912277498 0917955938 3871210005 8876668925
8448700470 7725524970 6044465212 7130404321 1826101035
9118647666 2963858495 0874484973 7347686142 0880529443

Figure 67. The First Illegal Prime (see page 238)

(hexadecimal), this prime (see Figure 67 on page 239) forms a gzip file of the original C-source code (without tables) that decrypts the DVD Movie encryption scheme (DeCSS). It was illegal to distribute this source code in the United States, so surely that made this number also illegal.

6796997961737181246...(2058 digits)...2008288306560595271
The largest known prime quadruple starts at 4104082046·4800#+5651. It was discovered by Norman Luhn in 2005.

3255876603648438217...(3139 digits)...7568434052915829849
Physicist David Broadhurst (1947–) showed that the following integral is a prime integer.

$$\frac{2^{903}5^{682}}{514269}\int_0^\infty \frac{x^{906}\sin(x\log 2)}{\sinh(\pi x/2)}\left(\frac{1}{\cosh(\pi x/5)}+8\sinh^2(\pi x/5)\right)dx$$

This type of integral can arise in the study of quantum field theory as well as the theory of multiple zeta functions.

1000000000000000000...(10000 digits)...0000000000000033603
$10^{9999}+33603$ is the smallest **gigantic prime**. It was verified in 2003 by Jens Franke, Thorsten Kleinjung, and Tobias Wirth with their distributed version of an ECPP program, which gave the largest ECPP proof at the time. [Luhn]

1750498405893918377...(20562 digits)...2218159886801379167
Mills proved there were numbers A for which $\lfloor A^{3^n}\rfloor$ is prime for all n. If we use the conjectured least such A, we get the **Mills' primes**. This prime is the eighth, and currently largest known, Mills' prime. It was proven prime by François Morain with ECPP in 2006, surpassing the previous ECPP record by over 5000 digits, and can be written as follows.

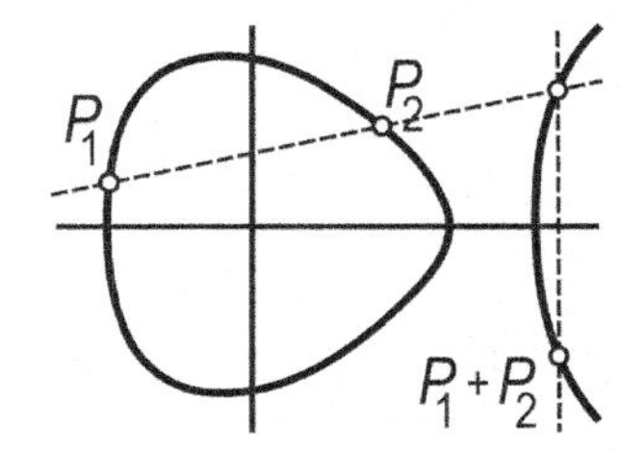

Figure 68. Elliptic Curve Addition (used by ECPP)

$$((((((2521008887^3+80)^3+12)^3+450)^3+894)^3+3636)^3+70756)^3+97220$$

The largest known **ordinary prime**.

11165936165811742\7...(100355 digits)...216443616890716159
The smaller member of the largest known pair of twin primes is $65516468355 \cdot 2^{333333} - 1$. It was discovered by "SG Grid" and others in August 2009, far surpassing the previous 58711-digit record of 2007!

58125794656333208 1...(142891 digits)...999999999999999999
The largest known factorial prime is $34790! - 1$. It was discovered by Marchal, Carmody, and Kuosa in 2002.

1698735164527416 2...(12837064 digits)...101954765562314751
The second largest known prime number ($2^{42643801} - 1$); first found by a machine on April 12, 2009. It is odd that no human took notice of this fact however until the following June 4! It is credited to Odd Magnar Strindmo from Melhus, Norway, who has been part of **GIMPS** since it began. This 47th known Mersenne prime is only 141125 digits (just over 1%) shorter than the largest known prime at that time.

3164702693302559 2...(12978189 digits)...022181166697152511
The largest known prime number ($2^{43112609} - 1$). It is the first known prime with 10 million or more digits and was discovered by Edson Smith of UCLA and GIMPS on August 23, 2008. This long-awaited prime was reported on a computer named `zeppelin.pic.ucla.edu`, a Dell OptiPlex 745 with an Intel® Core™2 Duo E6600 CPU running at 2.40 GHz.

The $100,000 Prime

Given the millennia that people have contemplated prime numbers, our continuing ignorance concerning the primes is stultifying.

Richard Crandall and Carl Pomerance

THE BRILLIANT deep blue Hope diamond is on permanent display at the Smithsonian Natural History Museum in Washington, D.C. Every year millions of admirers line up to see this beautiful and massive diamond. Diamonds this large are very rare, so they are extraordinarily valuable.

Perhaps the most valuable prime ever found was $2^{43112609}-1$. When Edson Smith found this mathematical gem in 2008, it was the first one ever found with more than ten million digits, so he won $100,000. He used software provided by the Great Internet Mersenne Prime Search (`GIMPS`), so he will share the prize with them. And what of his part? He will give that money to University of California at Los Angeles' (UCLA's) mathematics department. It was their computers he used to find the prime.

UCLA has been a rich source of record-sized primes—in fact, they have found the largest known prime eight times! In 1952, Professor Raphael Robinson found five Mersenne primes, and in 1961, Professor Alexander Hurwitz found two more.

Edson Smith's new prime was only the 45th Mersenne prime found in over 2000 years of searching, so like the Hope diamond, this prime is an exceedingly rare and beautiful find. And like diamonds, which we continue to mine, we know more primes will be found. In fact, before this one was verified, a second, slightly smaller Mersenne was found in Germany: $2^{37156667}-1$. This 11,185,272-digit prime became the 46th Mersenne prime found. A few months later, in April 2009, the 47th

Mersenne prime was found: $2^{42643801} - 1$. Just slightly smaller than the record we mentioned above, this 12,837,064-digit prime was found by Odd Magnar Strindmo of Norway.

The prize money came from the Electronic Frontier Foundation (EFF). There are still $400,000 in EFF funds waiting to be awarded for the first 100,000,000-digit prime and the first 1,000,000,000-digit prime. Perhaps you could win these prizes!

In this chapter we will very briefly discuss how these large primes are found. We will also survey the history (and future) of the record for the largest known prime.

Record Hand Calculations

In 1876, Édouard Lucas used a 127-by-127 "chessboard" as a computation aid to prove that $2^{127} - 1$, that is,

$$1701\ 41183\ 46046\ 92317\ 31687\ 30371\ 58841\ 05727,$$

is prime. This discovery should stand forever as the largest prime found by hand. Obviously he could not divide this jewel by every prime up to its square root! Instead, Lucas's greater discovery was a theorem that would later be simplified into the following indispensable tool for quarrying Mersenne primes.

Theorem (Lucas-Lehmer Test). *For odd primes p, the Mersenne number $2^p - 1$ is prime if and only if $2^p - 1$ divides S_{p-2} where $S_n = S_{n-1}^2 - 2$, and $S_0 = 4$.*

At about the same time Lucas made his discoveries, François Proth, a self-taught farmer, discovered the other key tool for excavating prime records.

Theorem (Proth's Theorem, 1878). *Let $n = h \cdot 2^k + 1$ with $2^k > h$. If there is an integer a such that $a^{(n-1)/2} \equiv -1 \pmod{n}$, then n is prime.*

All of the largest known primes found as of today use some form of these two theorems (see Table 21).

Computerized Records

Lucas's number held the record for 75 years, but then in June 1951 this record was broken twice. First, A. Ferrier used a mechanical desk

Table 21. The Ten Largest Known Primes

prime	digits	when	who & project
$2^{43112609}-1$	12978189	2008	Smith & GIMPS
$2^{42643801}-1$	12837064	2009	Strindmo & GIMPS
$2^{37156667}-1$	11185272	2008	Elvenich & GIMPS
$2^{32582657}-1$	9808358	2006	Cooper, Boone & GIMPS
$2^{30402457}-1$	9152052	2005	Cooper, Boone & GIMPS
$2^{25964951}-1$	7816230	2005	Nowak & GIMPS
$2^{24036583}-1$	7235733	2004	Findley & GIMPS
$2^{20996011}-1$	6320430	2003	Shafer & GIMPS
$2^{13466917}-1$	4053946	2001	Cameron, Kurowski & GIMPS
$19249 \cdot 2^{13018586}+1$	3918990	2007	Agafonov & Seventeen or Bust

calculator to extract the prime $(2^{148}+1)/17$, that is,

$$\begin{array}{r} 2098\backslash \\ 89366\ 57440\ 58648\ 61512\ 64256\ 61022\ 25938\ 63921. \end{array}$$

Next, Miller and Wheeler used the early electronic computer `EDSAC1` to unearth $180(2^{127}-1)^2+1$, that is,

$$\begin{array}{r} 5210\ 64401\ 56792\ 28794\ 06069\ 43253\ 90955\ 85333\backslash \\ 58984\ 83908\ 05645\ 83521\ 83851\ 01837\ 25557\ 35221. \end{array}$$

This record was soon eclipsed by Raphael Robinson's discoveries of five new Mersennes the very next year using the SWAC (Standards Western Automatic Computer) at UCLA. This was the first program that Robinson had ever written, and it ran the very first time he tried it. Not only that, but his program found two new record primes that very day! We see the records of Miller, Wheeler, and Robinson as the first points in Figure 69.

As large as these numbers seemed at the time, they are dwarfed by the 12,978,189-digit prime $2^{43112609}-1$ found last August by Edson Smith & `GIMPS`. In just seven score years the size of primes we can mine has grown from under 40 digits to over ten million digits! Yet, curiously, this exceptional growth is very orderly. In fact, the logarithm of the number of digits in the largest known prime has been surprisingly linear over the past 65 years ($R^2 = 0.9606$, see Figure 69).

Why is the graph in Figure 69 so linear? We will address that in the next section.

Figure 69. The Largest Known Prime by Year

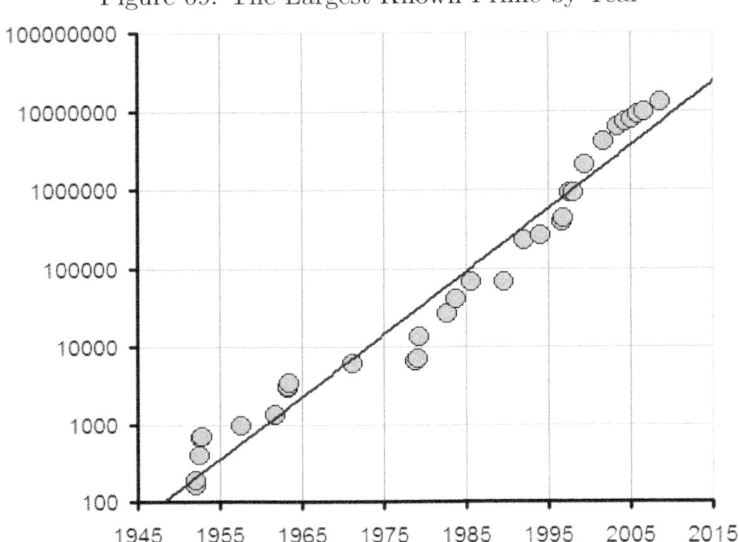

Moore's Law (almost)

In 1965, Gordon Moore, a cofounder of Intel, was asked to write an article predicting the development of semiconductor industry for the next decade. Moore used the data from the previous six years to predict that *the number of components* on the chips with the smallest manufacturing costs per component would double roughly every year. In 1975, he reduced the estimate to doubling every two years (the current rate seems to be closer to doubling every four years).

Later, an Intel colleague combined Moore's Law with the fact that clock speeds were increasing to estimate that computing power is doubling every 18 months. This power form of Moore's law may be the most common in circulation—even though Moore himself never used 18 months. More recent prognosticators have restated Moore's law in an economic form: the cost of computing power halves every x months.

The apparent linearity ($R^2 = 0.9606$) of the graph in Figure 69 implies a type of Moore's Law: the computing power devoted to finding large primes is increasing exponentially. The increase in power is coming from the increased transistor density, from faster clock speeds, from the aggregation of computing power into organized internet projects, as well as from the improved efficiency and accessibility of these projects.

Conjecture (Primal Moore's Law). *The quantity of computing power devoted to finding large primes is increasing exponentially.*

Notice this exponential growth holds at all levels of big prime hunting. Figure 70 shows the growth in the number of digits at several ranks in *The List of Largest Known Primes*—a database kept at `http://primes.utm.edu/primes`.

Figure 70. Digits in the nth Largest Prime

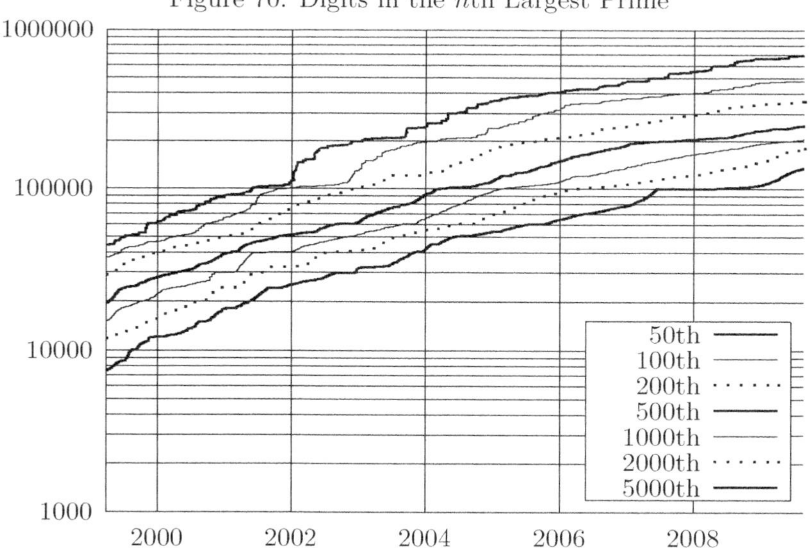

Quantification and Predictions

Suppose for a moment the computing power available for seeking primes doubles every k months, then one can show that the size of the largest known prime should double every $3k$ months. Currently, the line in Figure 69 (August 2008), after taking the log of the number of digits, has a slope of 0.079. This slope corresponds to doubling the digits every 3.8 years, or 46 months. So a quantified form of the primal Moore's law might be: the computing power available for seeking primes is doubling every 16 months.

How long can a primal Moore's law hold? Certainly as long as the economic form of Moore's law holds, and probably longer. Recently most of the computing power used to find large primes has been organized through massive projects involving tens of thousands of computers. For example, `GIMPS` currently involves over 75,000 computers. The growth and increased efficiency of these projects could extend the reign of the primal Moore's law. As computers become more ubiquitous and interconnected, perhaps someday even our toasters and refrigerators will be involved in this search for ever-larger primes.

The Electronic Frontier Foundation, `www.eff.org`, oversees prizes for the largest known primes. They have already paid $50,000 for the first 1,000,000-digit prime and $100,000 for the first prime with 10,000,000 digits. This leaves the $150,000 prize for the first 100,000,000-digit prime and the $250,000 prize for the first 1,000,000,000-digit prime yet to be claimed. At the current rate, we predict that primes with these sizes will be found in 2023 and 2035 respectively. (If you use only the primes found since 1975, the estimates are 2016 and 2025 respectively.) Perhaps just in time for the second and third editions of this book.

Will you be the one to unearth these precious jewels? The most active prime excavating projects can always be found at the web address below. Why not join us today?

`http://primes.utm.edu/bios/top20.php?type=project`

Appendices

Glossary

absolute prime a prime that remains prime for all permutations of its digits. E.g., 199, 919, and 991 are all primes. German mathematician H. E. Richert (1924–1993) originally called it a permutable prime.

almost-all-even-digits prime a prime with all even digits except for the rightmost digit. E.g., 86420864207 and 86420864209.

almost-equipandigital prime a prime with all digits (from 0 to 9) equal in number, except for one particular digit. E.g., $(1098765432123456789 0)_{42}1$.

alternate-digit prime a prime that has alternating odd and even digits. E.g., $(1676)_{948}1$.

balanced prime a prime number that is equal to the arithmetic mean of the nearest prime above and below. Algebraically speaking, given a prime number p_n, where n is its index in the ordered set of prime numbers, $p_n = (p_{n-1} + p_{n+1})/2$.

beastly prime a palindromic beastly prime has 666 in the center, 0's surrounding these digits, and 1 or 7 at the end. E.g., 700666007. A non-palindromic beastly prime begins with a 666, followed by 0's, with either a 1 or 7 at the right end. E.g., 6660000000001.

bemirp (or *bi*-directional *emirp*) a prime that yields a different prime when turned upside down with reversals of each being two more different primes. The result is four different associated primes. E.g., 1061 turned upside down yields 1901, and the reversals of these are 1601 and 1091 respectively.

Bertrand's postulate in 1845, French mathematician Joseph Bertrand (1822–1900) postulated that if $n > 3$, then there is at least one prime between n and $2n - 2$. It was proven by Pafnuty Lvovich Chebyshev (1821–1894) using elementary methods, and is therefore sometimes known as Chebyshev's theorem. In 1919, Ramanujan gave a simpler proof and a generalization, from which the concept of Ramanujan primes would later arise.

ceiling function denoted $\lceil x \rceil$, the least integer greater than or equal to x. $\lceil \pi \rceil = 4$, $\lceil -\pi \rceil = -3$, and $\lceil n \rceil = n$, for any integer n. The floor and ceiling functions appeared in Kenneth E. Iverson's *A Programming Language* (1962).

certificate of primality a short set of data that proves the primality of a number. It contains just enough information to reproduce the proof.

Chen prime a prime number p is called a Chen prime if $p + 2$ is either a prime or a semiprime. It is named after Chinese mathematician Chen Jingrun (1933–1996) who proved in 1966 that there are infinitely many such primes.

circular prime a prime that remains prime when we "rotate" its digits. E.g., 1193, 1931, 9311, and 3119 are primes. The known examples are 2, 3, 5, 7, 11, 13, 17, 37, 79, 113, 197, 199, 337, 1193, 3779, 11939, 19937, 193939, 199933, and of course the repunit primes greater than 11.

circular-digit prime a prime that has only the digits 0, 6, 8, or 9. E.g., 60686069. It is also called a loop-digit prime.

composite an integer *greater than one* which is not prime. Note that 1 is a unit, so is neither prime nor composite.

composite-digit prime a prime that has only the digits 4, 6, 8, or 9. E.g., 64486949.

congruence (or modular arithmetic) a is congruent to b modulo m, if m divides $a - b$. Gauss (1777–1855) introduced the notation $a \equiv b \pmod{m}$ in his pivotal book *Disquisitiones Arithmeticæ* (1801). E.g., $11 \equiv 3 \pmod{4}$.

congruent prime a prime that contains "shapes" of identical digits nested about the center when drawn in the form of symmetrical figures and polygons. The original prime can be constructed by concatenating each row left-to-right, top-to-bottom, as oriented in the drawing. E.g., 111181111 is a square-congruent prime.

coprime the integers a and b are said to be coprime (or relatively prime) if they have no common factor other than 1 or -1, or, equivalently, if their greatest common divisor is 1. E.g., 6 and 35 are coprime.

cousin primes a pair of prime numbers that differ by four. E.g., 3 and 7. There should be infinitely many of these.

cryptology the science of making and breaking secure codes. It consists of cryptography, the science of making secure codes, and cryptanalysis, the science of breaking them.

cuban prime a prime of the form $(n+1)^3 - n^3$. It is named after differences between successive cubic numbers. The sequence begins 7, 19, 37, 61, 127, 271, ...

Cullen prime a prime of the form $n \cdot 2^n + 1$ (the only known examples are n = 1, 141, 4713, 5795, 6611, 18496, 32292, 32469, 59656, 90825, 262419, 361275, 481899, 1354828, 6328548, and 6679881). It was first studied by Reverend James Cullen (1867–1933), an Irish Jesuit priest and schoolmaster. For Cullen prime of the 2nd kind, see Woodall prime.

Cunningham chain a Cunningham chain of length k (1st kind) is a sequence of k primes, each term of which is twice the preceding one plus 1. E.g., (2, 5, 11, 23, 47) and (89, 179, 359, 719, 1439, 2879). Note that a Cunningham chain of length k (2nd kind) is a sequence of k primes, each term of which is twice the preceding one minus 1. E.g., (2, 3, 5) and (1531, 3061, 6121, 12241, 24481).

curved-digit prime a prime that has only the digits 0, 3, 6, 8, or 9 (no straight lines). E.g., 83.

deletable prime a prime in which you can delete the digits one at a time in some order and get a prime at each step. It was first defined by Caldwell in 1987, a deletable prime year.

depression prime a palindromic prime having all equal internal digits and larger matching end digits. E.g., 101, 75557, and $7(2)_{723}7$.

dihedral prime a prime that remains prime when read on a 7-segment display (as seen on a digital clock or calculator) whether you hold it to a mirror, turn it upside down, or turn it upside down and hold it to a mirror. It must contain only the digits 0, 1, 2, 5, or 8. The sequence begins $2, 5, 11, 101, 181, 1181, 1811, \ldots$.

Dirichlet's theorem (on primes in arithmetic progressions) states that if a and b are relatively prime, then there are infinitely many primes of the form $an + b$ (like $3n + 1$ or $7n - 2$), for $n = 1, 2, 3, \ldots$.

divisor an integer d divides n if there is another integer q so that $dq = n$. In this case d is called a divisor (and a factor) of n and we denote this as $d|n$. Often the word is restricted to positive divisors, so the divisors of 6 are 1, 2, 3, and 6. (A divisor d of a number n is a unitary divisor if d and $\frac{n}{d}$ share no common factors.)

economical number one for which the prime factorization (including powers) requires fewer digits than the original number, such as $256 = 2^8$ (compare with extravagant number).

ECPP (acronym for Elliptic Curve Primality Proving) a class of algorithms that provide certificates of primality using sophisticated results from the theory of elliptic curves (see Figure 68, page 240). In practice, it is the fastest general-purpose primality-testing algorithm.

emirp (*prime* spelled backwards) a prime that gives a different prime when you reverse the order of its digits (compare with reversal). E.g., 389 and 983.

equidigital number one for which the prime factorization (including powers) requires the same number of digits as the original number. This includes all prime numbers.

Euler zeta function the following sum (defined for $\Re(s) > 1$) and equivalent product:

$$\zeta(s) = 1 + \frac{1}{2^s} + \frac{1}{3^s} + \frac{1}{4^s} + \ldots = \frac{1}{1-2^{-s}}\frac{1}{1-3^{-s}}\frac{1}{1-5^{-s}}\cdots.$$

extravagant number one for which the prime factorization (including powers) requires more digits than the original number, such as $30 = 2 \cdot 3 \cdot 5$ (compare with economical number).

factor see divisor.

factorial prime a prime of the form $n! \pm 1$. There are probably infinitely many, but only a few dozen are known.

Fermat number a number of the form $2^{2^n} + 1$. Fermat knew this is prime for $n = 0, 1, 2, 3$, and 4; but it now seems likely that all of the rest are composite.

Fermat prime a Fermat number that is prime. It is named after Pierre de Fermat (1601–1665), from a letter he wrote to Marin Mersenne (1588–1648) on December 25, 1640. The first few (and only known) are 3, 5, 17, 257, and 65537.

Fermat's little theorem states that if p is prime, then p divides $a^p - a$ for all integers a. When p does not divide a, this is sometimes written as $a^{p-1} \equiv 1 \pmod{p}$.

Fibonacci number a number in the sequence 1, 1, 2, 3, 5, 8, 13, 21, 34, 55, 89, ..., where each subsequent term is the sum of the preceding two. It was first described in the book *Liber Abaci* written by the Italian mathematician Leonardo of Pisa (Fibonacci) in 1202.

Fibonacci prime a prime Fibonacci number.

floor function denoted $\lfloor x \rfloor$, the greatest integer less than or equal to x. E.g., $\lfloor 5.8n \rfloor$ is prime for $n = 1, 2, \ldots, 5$.

Fortunate number let P be the product of the first n primes and let q be the least prime greater than $P + 1$. The nth Fortunate number is $q - P$. Reo F. Fortune (once married to the famed anthropologist Margaret Mead) conjectured that $q - P$ is always prime. The sequence begins $3, 5, 7, 13, 23, 17, 19, 23, \ldots$.

gaps (between primes) are runs of n consecutive composite numbers between two successive primes. The number of composites between one prime and the next is the length of the prime gap. E.g., there is a prime gap of length 7 between 89 and 97. Prime gaps have also been defined in terms of the differences between successive primes, which is one larger than this.

Gaussian Mersenne prime $(1 \pm i)^n - 1$ is a Gaussian Mersenne prime if and only if n is 2, or n is odd and the norm $2^n - (-1)^{(n^2-1)/8} 2^{(n+1)/2} + 1$ is a rational prime. This happens for $n = 2, 3, 5, 7, 11, 19, 29, 47, 73, 79, 113, 151, 157, 163, 167, \ldots$.

generalized Cullen prime any prime that can be written in the form $n \cdot b^n + 1$, with $n + 2 > b$. E.g., $669 \cdot 2^{128454} + 1$, which is $42816 \cdot 8^{42816} + 1$.

generalized Fermat prime a prime of the form $F_{b,n} = b^{2^n} + 1$, for an integer $b > 1$.

generalized repunit prime a prime that is a repunit in radix (base) b, i.e., $(b^n - 1)/(b - 1)$. Mersenne primes are generalized repunits in binary.

gigantic prime a prime with 10000 or more decimal digits. The term gigantic prime was coined by Samuel Yates.

GIMPS (an acronym for the Great Internet Mersenne Prime Search) a collaborative project seeking Mersenne primes (see `www.mersenne.org`).

Goldbach's conjecture (as reexpressed by Euler) asserts that every even integer $n > 2$ can be expressed as the sum of two primes. This conjecture was named after Prussian mathematician Christian Goldbach (1690–1764) who was the son of a pastor. The so-called Goldbach function, $G(n)$, gives the number of different ways that $2n$ can be expressed as the sum of two primes. A scatterplot known as "Goldbach's Comet" is a plot of this function (see Figure 48, page 149).

good prime a prime p_n is called "good" if $p_n^2 > p_{n-i} p_{n+i}$ for all $1 \leq i \leq n - 1$. The first few are 5, 11, 17, 29, 37, and 41.

high jumper see jumping champion.

holey prime (or wholly prime) a prime that has only digits with holes, i.e., 0, 4, 6, 8, or 9. E.g., 4649 and $9_{593}40004(9)_{593}$.

Honaker's problem asks for all consecutive prime number triples (p, q, r) with $p < q < r$ such that p divides $qr + 1$. It is likely that the only such triplets are $(2, 3, 5)$, $(3, 5, 7)$, and $(61, 67, 71)$.

iccanobiF prime a prime that becomes a Fibonacci number when reversed. E.g., 52057.

illegal prime a prime that represents information forbidden by law to possess or distribute. A prime found by Phil Carmody when written as a binary string was a computer program which bypasses copyright protection schemes on some DVD's. It was illegal at the time it was found.

integer the positive natural numbers: $1, 2, 3, \ldots$; their negatives: $-1, -2, -3, \ldots$; and zero.

invertible prime a prime that yields a different prime when the prime is inverted (turned upside down). E.g., 109 becomes 601. It must contain only the digits 0, 1, 6, 8, or 9. Charles W. Trigg (1898–1989) called the pair *prime rotative twins*.

irregular prime see regular prime.

jumping champion an integer n is called a jumping champion if n is the most frequently occurring difference between consecutive primes less than x, for some x. For $x = 3$, the jumping champion is 1; for $7 \leq x \leq 100$ it is 2; for $131 \leq x \leq 138$ it is 4; and for $389 \leq x \leq 420$ it is 6. Harry Nelson may have first suggested the concept in the late 1970's, but it was John H. Conway who coined the term jumping champion in 1993.

***k*-tuple** a prime k-tuple (also called a prime constellation) is a repeatable pattern of primes that are as close together as possible. E.g., twin primes are 2-tuples, prime triples are 3-tuples, prime quadruples are 4-tuples, etc.

left-truncatable prime (or simply truncatable prime) a prime number without the digit zero that remains prime no matter how many of the leading digits are omitted. E.g., 4632647 is left-truncatable because it and each of its truncations (632647,

32647, 2647, 647, 47, and 7) are primes. The digit zero is omitted to avoid trivial examples. It is also called a Russian doll prime.

Legendre's conjecture states that there is a prime number between n^2 and $(n+1)^2$ for every positive integer n. It is one of the four "unattackable" problems mentioned by Landau in the 1912 Fifth Congress of Mathematicians in Cambridge.

logarithm the power to which a base must be raised in order to obtain a given number. When the base $e = 2.7182818284...$ is used, it is called the natural logarithm. The function is denoted $\log x$, and sometimes $\ln x$. E.g., $\log 10 = 2.302585...$, because $e^{2.302585...} = 10$. See prime number theorem.

Lucas number a number in the Fibonacci-like sequence 2, 1, 3, 4, 7, 11, 18, 29, 47, ... It is named after French mathematician François Édouard Anatole Lucas (1842–1891).

Lucas prime a prime Lucas number.

megaprime a prime with 1,000,000 or more decimal digits.

Mersennary somebody who hunts for Mersenne primes only for the prize money.

Mersenne number an integer of the form $M(n) = 2^n - 1$. For the number to be prime it is necessary, but not sufficient, that n be prime.

Mersenne prime a prime Mersenne number ($M(p) = 2^p - 1$). It is named after the French monk Marin Mersenne (1588–1648) who communicated, in the preface to his *Cogitata Physico-Mathematica* (1644), some research about numbers of this form. Euclid had mentioned them in his ancient geometry book *Elements* almost two thousand years earlier.

Mills' prime in 1947, W. H. Mills proved there was a real constant A such that $\lfloor A^{3^n} \rfloor$ is prime for all positive integers n. The primes that the smallest choice of A gives are the Mills' primes.

minimal prime given any base b, there is a list of primes for which every prime of every length (when written in base b) has one of that list as a subsequence. (E.g., ABD is a subsequence of ABCDEF.) The smallest such list of primes are the minimal primes for base b. (The entries for 811 and 66600049 contain examples of minimal primes.)

multifactorial prime $n!_k = n(n-k)(n-2k) \cdot \ldots \cdot m$, where m is between 1 and k. Primes $n!_k \pm 1$ are multifactorial primes.

naughty prime a prime that is composed of mostly naughts (i.e., zeros). E.g., 1000303, $10^{24} + 7$, and $10^{60} + 7$.

near-repdigit prime a prime with all like or repeated digits but one. E.g., 7877 and 333337.

near-repunit prime a prime all but one of whose digits are 1. E.g., 1171 and 11111111113.

new Mersenne conjecture (or Bateman, Selfridge, and Wagstaff conjecture) states that for any odd natural number p, if any two of the following conditions hold, then so does the third: (i) $p = 2^k \pm 1$ or $p = 4^k \pm 3$ for some natural number k. (ii) $2^p - 1$ is a (Mersenne) prime. (iii) $(2^p + 1)/3$ is a (Wagstaff) prime. The conjecture can be thought of as an attempt to salvage the centuries old "Mersenne conjecture," which is false.

NSW prime a prime of the form $((1+\sqrt{2})^{2m+1} + (1-\sqrt{2})^{2m+1})/2$. The NSW stands for Newman, Shanks, and Williams (not New South Wales!)

number theory (or higher arithmetic) the study of the properties of integers. Gauss asserted that number theory is the "queen of mathematics."

ordinary prime a prime p for which none of $p^n \pm 1$ (for small n) factor enough to make the number easily provable using the classical methods of primality proof.

palindrome (from the Greek *palindromos*, "running back again") a word, verse, sentence, integer, etc., which reads the same forward or backward. E.g., racecar, or 31613.

palindromic prime (or palprime) a prime that is a palindrome. E.g., 133020331. A smoothly undulating palindromic prime (or SUPP) contains only two alternating digits. E.g., 74747474747474747, or any prime of the form $(ab)_n a$.

palindromic reflectable prime see triadic prime.

pandigital prime a prime with all 10 digits, i.e., from 0 to 9. The first few are 10123457689, 10123465789, and 10123465897.

perfect number a positive integer that equals the sum of its proper divisors, i.e., the sum of the positive divisors not including the number itself. The even perfect numbers, like 6 and 28, are a product of a Mersenne prime and a power of two. No odd perfect numbers are known. If odd perfect numbers do exist, then they are quite large (over 300 digits) and have numerous prime factors.

period (of a decimal expansion) the length of the repeating part (if any) of the decimal expansion of $\frac{1}{p}$. E.g., $\frac{1}{41} = 0.0243902439...$ has period 5. A full period prime (or long prime) is a prime p for which $\frac{1}{p}$ has the maximal period of $p-1$ digits.

Pierpont prime a prime having the form $2^u 3^v + 1$. The sequence begins $2, 3, 5, 7, 13, 17, 19, 37, 73, 97, 109, 163, \ldots$.

Pillai prime a prime number p for which there is an integer $n > 0$ such that $n!$ is one less than a multiple of p, while p is not one more than a multiple of n. It is named after Indian mathematician Subbayya Sivasankaranarayana Pillai (1901–1950) who proved that there are infinitely many such primes. The sequence begins $23, 29, 59, 61, 67, 71, \ldots$.

plateau prime a palindromic prime having all equal internal digits and smaller matching end digits. E.g., 181, 1777771, 355555553, and $5(10^{141} - 1)/9 - 2(10^{140} + 1)$.

prime counting function denoted $\pi(x)$, the number of primes less than or equal to x. E.g., $\pi(6521) = 843$. For large x, $\pi(x)$ is approximately $x/\log x$, where $\log x$ is the natural logarithm.

prime curiologist a person obsessed with "prime curios."

prime number an integer *greater than one* whose only positive divisors (factors) are 1 and itself. Note that 1 is neither prime nor composite, yet both a multiplicative identity and a unit. Euclid proved the number of primes is infinite in his *Elements* (Proposition 20, Book IX).

prime number theorem (or PNT) a theorem stating that the number of primes less than or equal to x, is about $x/\log x$. The primes "thin out" as one looks at larger and larger numbers. This discovery can be traced as far back as Gauss, at age 15 (circa 1792), but a rigorous mathematical proof wasn't provided until the French and Belgian mathematicians J. Hadamard and C. J. de la Vallée Poussin independently did so in 1896.

prime rotative twins see invertible prime.

prime-digit prime a prime that has only the digits 2, 3, 5, or 7. E.g., 2357.

primeval number a number that contains more primes than any smaller number. Here "contains" refers to those primes formed from rearranging subsets of the original number's digits. E.g., five primes can be formed from the digits of 107 (7, 17, 71, 107, and 701), and no other number less than 107 can produce an equal or higher quantity.

primorial (n-primorial) denoted $n\#$, the product of the primes less than or equal to n. E.g., $6\# = 5\# = 5 \cdot 3 \cdot 2 = 30$. The notation was introduced by Harvey Dubner.

primorial prime a prime of the form $n\# + 1$ or $n\# - 1$ (see primorial). E.g., $31\# + 1 = 200560490131$ is prime.

probable prime a number that passes a test also passed by all the primes, but passed by few composites. The base a Fermat probable primes (a-PRP's) are those n (greater than 1 and coprime to a) which divide $a^{n-1} - 1$.

proper divisor (or aliquot divisor) any positive divisor of n other than n itself. E.g., the proper divisors of 6 are 1, 2, and 3.

Proth prime a prime of the form $k \cdot 2^n + 1$ with k odd and $2^n > k$. It is named after the self-taught mathematician farmer François Proth (1852–1879) who lived near Verdun, France, and published a theorem for proving the number's primality.

pseudoprime a composite probable prime. At one time all probable primes were called pseudoprimes, but now this term is limited to composites.

public-key cryptography a type of cryptography in which the encoding key is revealed without compromising the encoded message. E.g., the RSA algorithm.

Ramanujan prime one of the integers R_n that are the smallest to satisfy the condition $\pi(x) - \pi(\frac{x}{2}) \geq n$, for all $x \geq R_n$. In other words, there are at least n primes between $\frac{x}{2}$ and x, whenever x is greater than or equal to R_n. The sequence begins $2, 11, 17, 29, 41, \ldots$.

reflectable prime a prime that is invariant upon mirror reflection along the line its digits are written on. It must contain only the digits 0, 1, 3, or 8. E.g., 181031.

regular prime an odd prime number p is regular if it does not divide the class number of the pth cyclotomic field (obtained by adjoining a primitive pth root of unity to the field of rationals); otherwise it is irregular. Ernst Kummer (1810–1893) established equivalently that an odd prime p is regular if (and only if) it does not divide the numerator of any of the Bernoulli numbers B_k (see 283 on page 84), for $k = 2, 4, 6, \ldots, p-3$.

repunit prime a prime whose digits are all 1's. The repunits are defined as $R_n = (10^n - 1)/9$ for $n \geq 1$, where the number R_n consists of n copies of the digit 1. E.g., 11, 1111111111111111111, and 11111111111111111111111.

reversal the reversal of a number *abc...* is *...cba*. A reversible prime is a prime that remains prime when its digits are reversed. Reversible primes can be divided into two sets: emirps and palindromic primes.

Riemann hypothesis states that the real part of any non-trivial zero (solution) of the Riemann zeta function is $\frac{1}{2}$. It has remained unproven ever since its formulation by Bernhard Riemann in 1859, and is central to understanding the general distribution of primes. A $1,000,000 prize has been offered by the Clay Mathematics Institute for a proof.

Riemann zeta function Riemann extended Euler's zeta function to one defined for all complex numbers except 1, and which obeys the functional equation:

$$\pi^{-s/2}\Gamma\left(\frac{s}{2}\right)\zeta(s) = \pi^{-(1-s)/2}\Gamma\left(\frac{1-s}{2}\right)\zeta(1-s).$$

right-truncatable prime (or snowball prime) a prime that remains prime even if you stop before writing all of the digits. E.g., 73939133.

RSA algorithm perhaps the most famous of all public-key cryptosystems. Ronald Rivest, Adi Shamir, and Leonard Adleman at MIT announced it in 1977. It relies on the relative ease of finding large primes and the comparative difficulty of factoring integers for its security.

RSA numbers are composite numbers having exactly two prime factors (semiprimes) that have been listed in the factoring challenges conducted by RSA Laboratories.

safe prime a prime p for which $(p-1)/2$ is also prime. It is "safer" when used in certain types of encryption.

self-descriptive prime a prime that describes itself as the digits are read in pairs. E.g., 10153331 reads "one 0, one 5, three 3's, and three 1's."

semiprime the product of two primes, which is sometimes called a $P2$ or a 2-almost prime. The largest known semiprime is always the square of the largest known prime.

sexy primes a pair of prime numbers that differ by six. E.g., 5 and 11. These are so-named because *sex* is the Latin word for six.

Sieve of Eratosthenes a simple, ancient algorithm for finding all prime numbers up to a specified integer via crossing-out numbers known to be composite in an orderly way (see Figure 50). Eratosthenes of Cyrene (now Shahhat, Libya) is credited with discovering this method, circa 240 B.C. It is the predecessor to the modern Sieve of Atkin, which is faster but more complex.

Smarandache-Wellin prime a prime that is the concatenation of the first n prime numbers. E.g., 2, 23, 2357, and the concatenation of the first 128 primes. It is also known as a concatenate prime.

smoothly undulating see palindromic prime.

snowball prime see right-truncatable prime.

Sophie Germain prime a prime p for which $2p + 1$ is also prime (compare with safe prime). Around 1825, Sophie Germain proved that the first case of Fermat's last theorem is true for odd Germain primes.

straight-digit prime a prime that has only the digits 1, 4, or 7 (no curves). E.g., 1447.

strobogrammatic a number is strobogrammatic if it remains the same when turned upside down (see Figure 71 on page 265). It must contain only the digits 0, 1, 6, 8, or 9. E.g., 619. Note that strobogrammatic numbers on a 7-segment display can also contain the digits 2 and 5.

tetradic prime a prime which is the same forward and backward (palindromic) as well as upside down and mirror reflected along the line its digits are written on. It must contain only the digits 0, 1, or 8. E.g., 18181. Tetradic primes are also known as 4-way primes.

titanic prime a prime with 1000 or more decimal digits. In 1984, Samuel Yates coined the name and called those who proved their primality "titans."

triadic prime a prime which is the same forward and backward (palindromic) as well as mirror reflected only along the line its digits are written on. It must contain only the digits 0, 1, 3, or 8. E.g., 131. Charles W. Trigg called this a palindromic reflectable prime. Triadic primes are also known as 3-way primes.

truncatable prime see left-truncatable prime.

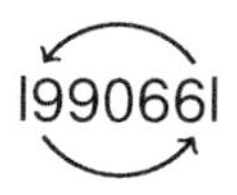

Figure 71. Turning a Number Upside Down

twin primes a pair of prime numbers that differ by two. E.g., 3 and 5. The term "twin prime" was coined by Paul Stäckel in 1916. It is conjectured that there are infinitely many of these.

unholey prime a prime that does not have any digits with holes in them (see holey prime).

unique prime (or unique period prime) a prime p (other than 2 and 5) which has a period (i.e., the decimal expansion of $\frac{1}{p}$ repeats in blocks of some set length) that it shares with no other prime. The period of a prime p always divides $p-1$. It was defined by Samuel Yates in 1980.

upside down refers to the process of inverting a number's decimal representation by rotating it 180° (see the counter-clockwise example in Figure 71).

Vinogradov's theorem states that every sufficiently large odd integer is a sum of at most 3 primes. It is closely related to both Goldbach's and Waring's prime number conjectures and named after Russian mathematician Ivan Matveyevich Vinogradov (1891–1983).

Wagstaff prime a prime of the form $(2^p+1)/3$. The number appears in the new Mersenne conjecture and has applications in cryptography. It is named after Samuel S. Wagstaff, Jr., a professor of computer science at Purdue University.

Wall-Sun-Sun prime a prime $p > 5$ such that p^2 divides the Fibonacci number fib$(p - (p|5))$, where $(p|5)$ is a Legendre symbol. None are known! It is named after D. D. Wall, Zhi-Hong Sun, and his twin brother Zhi-Wei Sun.

weakly prime a prime is said to be "weakly" if changing a single digit to every other possible digit produces a composite number when performed on each digit.

Wieferich prime a prime p such that p^2 divides $2^{p-1} - 1$. Note that 1093 and 3511 are the only known examples. All odd primes p divide $2^{p-1} - 1$.

Wilson prime a prime p such that p^2 divides $(p-1)! + 1$. The only known Wilson primes are 5, 13, and 563. There are no others less than 500,000,000. All primes p divide $(p-1)! + 1$.

Wolstenholme prime a prime p which divides B_{p-3}, where B_n is the nth Bernoulli number, or equivalently, a prime p for which the central binomial coefficient $\binom{2p}{p} \equiv 2 \pmod{p^4}$. After searching through all primes up to 1,000,000,000, the only known Wolstenholme primes remain the lonely pair 16843 and 2124679. Heuristically speaking, there should be roughly one Wolstenholme prime between 10^9 and 10^{24}.

Woodall prime a prime of the form $n \cdot 2^n - 1$. A Woodall prime is sometimes called a Cullen prime of the 2nd kind.

Yarborough prime a prime that does not contain the digits 0 and 1. An anti-Yarborough prime contains only 0's and 1's.

zeta function see Euler zeta function and Riemann zeta function.

Prime Sites

BIGprimes.net
`http://www.bigprimes.net/`
An online archive of prime numbers with a built-in 'number cruncher.'

The Cunningham Project
`http://homes.cerias.purdue.edu/~ssw/cun/`
Factors of the numbers $b^n \pm 1$ for small integers b.

Michael Hartley's Maths Page
`http://www.dr-mikes-maths.com/maths.html`
A wonderful math site which includes Ulam Prime Spirals and a search for large primes of the form $k \cdot 2^n - 1$.

The Great Internet Mersenne Prime Search (GIMPS)
`http://www.mersenne.org/`
This project provides the free software that has found *all* of the recent record Mersenne primes.

The Math Forum @Drexel
`http://mathforum.org/`
Comprehensive resource for K-12 educators.

The Mathematical Association of America (MAA)
`http://www.maa.org/`
Organization of two and four-year college professors.

Mudd Math Fun Facts
`http://www.math.hmc.edu/funfacts/`
A resource for enriching your math courses and nurturing your interest and talent in mathematics.

National Council of Teachers of Mathematics (NCTM)

`http://www.nctm.org/`

K-12 math resources and development opportunities.

Number Recreations

`http://www.shyamsundergupta.com/`

Recreational topics that range from prime polynomials to pseudoprime curiosities.

The On-Line Encyclopedia of Integer Sequences (OEIS)

`http://www.research.att.com/~njas/sequences/`

Enter a few terms of your sequence and see what is known.

Patterns in Primes

`http://www.geocities.com/~harveyh/primes.htm`

Palindromes, reversible primes, and so much more!

The Prime Pages

`http://primes.utm.edu`

Prime number research, records, and results. The database of the 5000 largest known primes is updated hourly.

The Prime Puzzles and Problems Connection

`http://www.primepuzzles.net/`

A well presented anthology of the interesting problems and puzzles explicitly related to primes.

Primes in Arithmetic Progression Records

`http://users.cybercity.dk/~ds1522332/math/aprecords.htm`

Records carefully collected and actively updated.

Wolfram's MathWorld

`http://mathworld.wolfram.com/`

An extensive encyclopedia of mathematics with a substantial amount of information on primes.

World!Of Numbers

`http://www.worldofnumbers.com/`

This excellent website presents a variety of recreational topics and includes a special section on palindromic primes.

Prime Books

CRC Concise Encyclopedia of Mathematics by E. Weisstein, 2nd edition, Chapman & Hall/CRC, 2003; ISBN 1-584-88347-2.

Édouard Lucas and Primality Testing by H. C. Williams, Canadian Mathematical Society Series of Monographs and Advanced Texts, volume 22, John Wiley & Sons, New York, 1998; ISBN 0-471-14852-0.

Elementary Number Theory, by G. A. Jones and J. M. Jones, Springer, 2009 (corrected edition); ISBN 3-5407-6197-7.

Elementary Number Theory, 6th Edition, by D. M. Burton, McGraw Hill, 2005; ISBN 0-0706-1607-8.

Exploring Prime Numbers on Your PC and the Internet by E. Haga, Enoch Haga Publisher, Folsom, California, First Revised Edition 2007; ISBN 978-1-885794-24-6.

The Kingdom of the Infinite Number–A Field Guide by B. Bunch, W. H. Freeman and Company, 2001; ISBN 0-7167-4447-3.

The Little Book of Bigger Primes by P. Ribenboim, 2nd edition, Springer-Verlag, New York, 2004; ISBN 0-387-20169-6.

Lure of the Integers by J. Roberts, Mathematical Association of America, 1992; ISBN 0-88385-502-X.

Merveilleux nombres premiers: Voyage au cœur de l'arithmétique by J. Delahaye, Éditions Belin/Pour la science, Paris, 2000; ISBN 2-84245-017-5.

The Music of the Primes: Searching to Solve the Greatest Mystery in Mathematics by M. du Sautoy, Perennial Edition, 2004; ISBN 0-06-093558-8.

The Penguin Book of Curious and Interesting Numbers: Revised Edition by D. Wells, Penguin Press Science, 1998; ISBN 0-140-26149-4.

Prime Numbers: A Computational Perspective by R. Crandall and C. Pomerance, 2nd edition, Springer-Verlag, New York, 2005; ISBN 978-0-387-25282-7.

Prime Numbers: The Most Mysterious Figures in Mathematics by D. Wells, John Wiley & Sons, Inc., 2005; ISBN 0-471-46234-9.

Prime Obsession: Bernhard Riemann and the Greatest Unsolved Problem in Mathematics by J. Derbyshire, Plume, 2004; ISBN 0-452-28525-9.

Those Fascinating Numbers by J. M. De Koninck, American Mathematical Society, 2009; ISBN 0-8218-4807-0.

Unsolved Problems in Number Theory by R. Guy, 3rd edition, Springer-Verlag, New York, 2004; ISBN 0-387-20860-7.

Table 22. The Primes Less Than $\sqrt{10^9}$

2	3	5	7	11	13	17	19	23	29
31	37	41	43	47	53	59	61	67	71
73	79	83	89	97	101	103	107	109	113
127	131	137	139	149	151	157	163	167	173
179	181	191	193	197	199	211	223	227	229
233	239	241	251	257	263	269	271	277	281
283	293	307	311	313	317	331	337	347	349
353	359	367	373	379	383	389	397	401	409
419	421	431	433	439	443	449	457	461	463
467	479	487	491	499	503	509	521	523	541
547	557	563	569	571	577	587	593	599	601
607	613	617	619	631	641	643	647	653	659
661	673	677	683	691	701	709	719	727	733
739	743	751	757	761	769	773	787	797	809
811	821	823	827	829	839	853	857	859	863
877	881	883	887	907	911	919	929	937	941
947	953	967	971	977	983	991	997	1009	1013
1019	1021	1031	1033	1039	1049	1051	1061	1063	1069
1087	1091	1093	1097	1103	1109	1117	1123	1129	1151
1153	1163	1171	1181	1187	1193	1201	1213	1217	1223
1229	1231	1237	1249	1259	1277	1279	1283	1289	1291
1297	1301	1303	1307	1319	1321	1327	1361	1367	1373
1381	1399	1409	1423	1427	1429	1433	1439	1447	1451
1453	1459	1471	1481	1483	1487	1489	1493	1499	1511
1523	1531	1543	1549	1553	1559	1567	1571	1579	1583
1597	1601	1607	1609	1613	1619	1621	1627	1637	1657
1663	1667	1669	1693	1697	1699	1709	1721	1723	1733
1741	1747	1753	1759	1777	1783	1787	1789	1801	1811
1823	1831	1847	1861	1867	1871	1873	1877	1879	1889
1901	1907	1913	1931	1933	1949	1951	1973	1979	1987
1993	1997	1999	2003	2011	2017	2027	2029	2039	2053
2063	2069	2081	2083	2087	2089	2099	2111	2113	2129
2131	2137	2141	2143	2153	2161	2179	2203	2207	2213
2221	2237	2239	2243	2251	2267	2269	2273	2281	2287
2293	2297	2309	2311	2333	2339	2341	2347	2351	2357

2371	2377	2381	2383	2389	2393	2399	2411	2417	2423
2437	2441	2447	2459	2467	2473	2477	2503	2521	2531
2539	2543	2549	2551	2557	2579	2591	2593	2609	2617
2621	2633	2647	2657	2659	2663	2671	2677	2683	2687
2689	2693	2699	2707	2711	2713	2719	2729	2731	2741
2749	2753	2767	2777	2789	2791	2797	2801	2803	2819
2833	2837	2843	2851	2857	2861	2879	2887	2897	2903
2909	2917	2927	2939	2953	2957	2963	2969	2971	2999
3001	3011	3019	3023	3037	3041	3049	3061	3067	3079
3083	3089	3109	3119	3121	3137	3163	3167	3169	3181
3187	3191	3203	3209	3217	3221	3229	3251	3253	3257
3259	3271	3299	3301	3307	3313	3319	3323	3329	3331
3343	3347	3359	3361	3371	3373	3389	3391	3407	3413
3433	3449	3457	3461	3463	3467	3469	3491	3499	3511
3517	3527	3529	3533	3539	3541	3547	3557	3559	3571
3581	3583	3593	3607	3613	3617	3623	3631	3637	3643
3659	3671	3673	3677	3691	3697	3701	3709	3719	3727
3733	3739	3761	3767	3769	3779	3793	3797	3803	3821
3823	3833	3847	3851	3853	3863	3877	3881	3889	3907
3911	3917	3919	3923	3929	3931	3943	3947	3967	3989
4001	4003	4007	4013	4019	4021	4027	4049	4051	4057
4073	4079	4091	4093	4099	4111	4127	4129	4133	4139
4153	4157	4159	4177	4201	4211	4217	4219	4229	4231
4241	4243	4253	4259	4261	4271	4273	4283	4289	4297
4327	4337	4339	4349	4357	4363	4373	4391	4397	4409
4421	4423	4441	4447	4451	4457	4463	4481	4483	4493
4507	4513	4517	4519	4523	4547	4549	4561	4567	4583
4591	4597	4603	4621	4637	4639	4643	4649	4651	4657
4663	4673	4679	4691	4703	4721	4723	4729	4733	4751
4759	4783	4787	4789	4793	4799	4801	4813	4817	4831
4861	4871	4877	4889	4903	4909	4919	4931	4933	4937
4943	4951	4957	4967	4969	4973	4987	4993	4999	5003
5009	5011	5021	5023	5039	5051	5059	5077	5081	5087
5099	5101	5107	5113	5119	5147	5153	5167	5171	5179
5189	5197	5209	5227	5231	5233	5237	5261	5273	5279
5281	5297	5303	5309	5323	5333	5347	5351	5381	5387
5393	5399	5407	5413	5417	5419	5431	5437	5441	5443
5449	5471	5477	5479	5483	5501	5503	5507	5519	5521
5527	5531	5557	5563	5569	5573	5581	5591	5623	5639
5641	5647	5651	5653	5657	5659	5669	5683	5689	5693
5701	5711	5717	5737	5741	5743	5749	5779	5783	5791
5801	5807	5813	5821	5827	5839	5843	5849	5851	5857
5861	5867	5869	5879	5881	5897	5903	5923	5927	5939
5953	5981	5987	6007	6011	6029	6037	6043	6047	6053

6067	6073	6079	6089	6091	6101	6113	6121	6131	6133
6143	6151	6163	6173	6197	6199	6203	6211	6217	6221
6229	6247	6257	6263	6269	6271	6277	6287	6299	6301
6311	6317	6323	6329	6337	6343	6353	6359	6361	6367
6373	6379	6389	6397	6421	6427	6449	6451	6469	6473
6481	6491	6521	6529	6547	6551	6553	6563	6569	6571
6577	6581	6599	6607	6619	6637	6653	6659	6661	6673
6679	6689	6691	6701	6703	6709	6719	6733	6737	6761
6763	6779	6781	6791	6793	6803	6823	6827	6829	6833
6841	6857	6863	6869	6871	6883	6899	6907	6911	6917
6947	6949	6959	6961	6967	6971	6977	6983	6991	6997
7001	7013	7019	7027	7039	7043	7057	7069	7079	7103
7109	7121	7127	7129	7151	7159	7177	7187	7193	7207
7211	7213	7219	7229	7237	7243	7247	7253	7283	7297
7307	7309	7321	7331	7333	7349	7351	7369	7393	7411
7417	7433	7451	7457	7459	7477	7481	7487	7489	7499
7507	7517	7523	7529	7537	7541	7547	7549	7559	7561
7573	7577	7583	7589	7591	7603	7607	7621	7639	7643
7649	7669	7673	7681	7687	7691	7699	7703	7717	7723
7727	7741	7753	7757	7759	7789	7793	7817	7823	7829
7841	7853	7867	7873	7877	7879	7883	7901	7907	7919
7927	7933	7937	7949	7951	7963	7993	8009	8011	8017
8039	8053	8059	8069	8081	8087	8089	8093	8101	8111
8117	8123	8147	8161	8167	8171	8179	8191	8209	8219
8221	8231	8233	8237	8243	8263	8269	8273	8287	8291
8293	8297	8311	8317	8329	8353	8363	8369	8377	8387
8389	8419	8423	8429	8431	8443	8447	8461	8467	8501
8513	8521	8527	8537	8539	8543	8563	8573	8581	8597
8599	8609	8623	8627	8629	8641	8647	8663	8669	8677
8681	8689	8693	8699	8707	8713	8719	8731	8737	8741
8747	8753	8761	8779	8783	8803	8807	8819	8821	8831
8837	8839	8849	8861	8863	8867	8887	8893	8923	8929
8933	8941	8951	8963	8969	8971	8999	9001	9007	9011
9013	9029	9041	9043	9049	9059	9067	9091	9103	9109
9127	9133	9137	9151	9157	9161	9173	9181	9187	9199
9203	9209	9221	9227	9239	9241	9257	9277	9281	9283
9293	9311	9319	9323	9337	9341	9343	9349	9371	9377
9391	9397	9403	9413	9419	9421	9431	9433	9437	9439
9461	9463	9467	9473	9479	9491	9497	9511	9521	9533
9539	9547	9551	9587	9601	9613	9619	9623	9629	9631
9643	9649	9661	9677	9679	9689	9697	9719	9721	9733
9739	9743	9749	9767	9769	9781	9787	9791	9803	9811
9817	9829	9833	9839	9851	9857	9859	9871	9883	9887
9901	9907	9923	9929	9931	9941	9949	9967	9973	10007

10009	10037	10039	10061	10067	10069	10079	10091	10093	10099
10103	10111	10133	10139	10141	10151	10159	10163	10169	10177
10181	10193	10211	10223	10243	10247	10253	10259	10267	10271
10273	10289	10301	10303	10313	10321	10331	10333	10337	10343
10357	10369	10391	10399	10427	10429	10433	10453	10457	10459
10463	10477	10487	10499	10501	10513	10529	10531	10559	10567
10589	10597	10601	10607	10613	10627	10631	10639	10651	10657
10663	10667	10687	10691	10709	10711	10723	10729	10733	10739
10753	10771	10781	10789	10799	10831	10837	10847	10853	10859
10861	10867	10883	10889	10891	10903	10909	10937	10939	10949
10957	10973	10979	10987	10993	11003	11027	11047	11057	11059
11069	11071	11083	11087	11093	11113	11117	11119	11131	11149
11159	11161	11171	11173	11177	11197	11213	11239	11243	11251
11257	11261	11273	11279	11287	11299	11311	11317	11321	11329
11351	11353	11369	11383	11393	11399	11411	11423	11437	11443
11447	11467	11471	11483	11489	11491	11497	11503	11519	11527
11549	11551	11579	11587	11593	11597	11617	11621	11633	11657
11677	11681	11689	11699	11701	11717	11719	11731	11743	11777
11779	11783	11789	11801	11807	11813	11821	11827	11831	11833
11839	11863	11867	11887	11897	11903	11909	11923	11927	11933
11939	11941	11953	11959	11969	11971	11981	11987	12007	12011
12037	12041	12043	12049	12071	12073	12097	12101	12107	12109
12113	12119	12143	12149	12157	12161	12163	12197	12203	12211
12227	12239	12241	12251	12253	12263	12269	12277	12281	12289
12301	12323	12329	12343	12347	12373	12377	12379	12391	12401
12409	12413	12421	12433	12437	12451	12457	12473	12479	12487
12491	12497	12503	12511	12517	12527	12539	12541	12547	12553
12569	12577	12583	12589	12601	12611	12613	12619	12637	12641
12647	12653	12659	12671	12689	12697	12703	12713	12721	12739
12743	12757	12763	12781	12791	12799	12809	12821	12823	12829
12841	12853	12889	12893	12899	12907	12911	12917	12919	12923
12941	12953	12959	12967	12973	12979	12983	13001	13003	13007
13009	13033	13037	13043	13049	13063	13093	13099	13103	13109
13121	13127	13147	13151	13159	13163	13171	13177	13183	13187
13217	13219	13229	13241	13249	13259	13267	13291	13297	13309
13313	13327	13331	13337	13339	13367	13381	13397	13399	13411
13417	13421	13441	13451	13457	13463	13469	13477	13487	13499
13513	13523	13537	13553	13567	13577	13591	13597	13613	13619
13627	13633	13649	13669	13679	13681	13687	13691	13693	13697
13709	13711	13721	13723	13729	13751	13757	13759	13763	13781
13789	13799	13807	13829	13831	13841	13859	13873	13877	13879
13883	13901	13903	13907	13913	13921	13931	13933	13963	13967
13997	13999	14009	14011	14029	14033	14051	14057	14071	14081
14083	14087	14107	14143	14149	14153	14159	14173	14177	14197

14207	14221	14243	14249	14251	14281	14293	14303	14321	14323
14327	14341	14347	14369	14387	14389	14401	14407	14411	14419
14423	14431	14437	14447	14449	14461	14479	14489	14503	14519
14533	14537	14543	14549	14551	14557	14561	14563	14591	14593
14621	14627	14629	14633	14639	14653	14657	14669	14683	14699
14713	14717	14723	14731	14737	14741	14747	14753	14759	14767
14771	14779	14783	14797	14813	14821	14827	14831	14843	14851
14867	14869	14879	14887	14891	14897	14923	14929	14939	14947
14951	14957	14969	14983	15013	15017	15031	15053	15061	15073
15077	15083	15091	15101	15107	15121	15131	15137	15139	15149
15161	15173	15187	15193	15199	15217	15227	15233	15241	15259
15263	15269	15271	15277	15287	15289	15299	15307	15313	15319
15329	15331	15349	15359	15361	15373	15377	15383	15391	15401
15413	15427	15439	15443	15451	15461	15467	15473	15493	15497
15511	15527	15541	15551	15559	15569	15581	15583	15601	15607
15619	15629	15641	15643	15647	15649	15661	15667	15671	15679
15683	15727	15731	15733	15737	15739	15749	15761	15767	15773
15787	15791	15797	15803	15809	15817	15823	15859	15877	15881
15887	15889	15901	15907	15913	15919	15923	15937	15959	15971
15973	15991	16001	16007	16033	16057	16061	16063	16067	16069
16073	16087	16091	16097	16103	16111	16127	16139	16141	16183
16187	16189	16193	16217	16223	16229	16231	16249	16253	16267
16273	16301	16319	16333	16339	16349	16361	16363	16369	16381
16411	16417	16421	16427	16433	16447	16451	16453	16477	16481
16487	16493	16519	16529	16547	16553	16561	16567	16573	16603
16607	16619	16631	16633	16649	16651	16657	16661	16673	16691
16693	16699	16703	16729	16741	16747	16759	16763	16787	16811
16823	16829	16831	16843	16871	16879	16883	16889	16901	16903
16921	16927	16931	16937	16943	16963	16979	16981	16987	16993
17011	17021	17027	17029	17033	17041	17047	17053	17077	17093
17099	17107	17117	17123	17137	17159	17167	17183	17189	17191
17203	17207	17209	17231	17239	17257	17291	17293	17299	17317
17321	17327	17333	17341	17351	17359	17377	17383	17387	17389
17393	17401	17417	17419	17431	17443	17449	17467	17471	17477
17483	17489	17491	17497	17509	17519	17539	17551	17569	17573
17579	17581	17597	17599	17609	17623	17627	17657	17659	17669
17681	17683	17707	17713	17729	17737	17747	17749	17761	17783
17789	17791	17807	17827	17837	17839	17851	17863	17881	17891
17903	17909	17911	17921	17923	17929	17939	17957	17959	17971
17977	17981	17987	17989	18013	18041	18043	18047	18049	18059
18061	18077	18089	18097	18119	18121	18127	18131	18133	18143
18149	18169	18181	18191	18199	18211	18217	18223	18229	18233
18251	18253	18257	18269	18287	18289	18301	18307	18311	18313
18329	18341	18353	18367	18371	18379	18397	18401	18413	18427

18433	18439	18443	18451	18457	18461	18481	18493	18503	18517
18521	18523	18539	18541	18553	18583	18587	18593	18617	18637
18661	18671	18679	18691	18701	18713	18719	18731	18743	18749
18757	18773	18787	18793	18797	18803	18839	18859	18869	18899
18911	18913	18917	18919	18947	18959	18973	18979	19001	19009
19013	19031	19037	19051	19069	19073	19079	19081	19087	19121
19139	19141	19157	19163	19181	19183	19207	19211	19213	19219
19231	19237	19249	19259	19267	19273	19289	19301	19309	19319
19333	19373	19379	19381	19387	19391	19403	19417	19421	19423
19427	19429	19433	19441	19447	19457	19463	19469	19471	19477
19483	19489	19501	19507	19531	19541	19543	19553	19559	19571
19577	19583	19597	19603	19609	19661	19681	19687	19697	19699
19709	19717	19727	19739	19751	19753	19759	19763	19777	19793
19801	19813	19819	19841	19843	19853	19861	19867	19889	19891
19913	19919	19927	19937	19949	19961	19963	19973	19979	19991
19993	19997	20011	20021	20023	20029	20047	20051	20063	20071
20089	20101	20107	20113	20117	20123	20129	20143	20147	20149
20161	20173	20177	20183	20201	20219	20231	20233	20249	20261
20269	20287	20297	20323	20327	20333	20341	20347	20353	20357
20359	20369	20389	20393	20399	20407	20411	20431	20441	20443
20477	20479	20483	20507	20509	20521	20533	20543	20549	20551
20563	20593	20599	20611	20627	20639	20641	20663	20681	20693
20707	20717	20719	20731	20743	20747	20749	20753	20759	20771
20773	20789	20807	20809	20849	20857	20873	20879	20887	20897
20899	20903	20921	20929	20939	20947	20959	20963	20981	20983
21001	21011	21013	21017	21019	21023	21031	21059	21061	21067
21089	21101	21107	21121	21139	21143	21149	21157	21163	21169
21179	21187	21191	21193	21211	21221	21227	21247	21269	21277
21283	21313	21317	21319	21323	21341	21347	21377	21379	21383
21391	21397	21401	21407	21419	21433	21467	21481	21487	21491
21493	21499	21503	21517	21521	21523	21529	21557	21559	21563
21569	21577	21587	21589	21599	21601	21611	21613	21617	21647
21649	21661	21673	21683	21701	21713	21727	21737	21739	21751
21757	21767	21773	21787	21799	21803	21817	21821	21839	21841
21851	21859	21863	21871	21881	21893	21911	21929	21937	21943
21961	21977	21991	21997	22003	22013	22027	22031	22037	22039
22051	22063	22067	22073	22079	22091	22093	22109	22111	22123
22129	22133	22147	22153	22157	22159	22171	22189	22193	22229
22247	22259	22271	22273	22277	22279	22283	22291	22303	22307
22343	22349	22367	22369	22381	22391	22397	22409	22433	22441
22447	22453	22469	22481	22483	22501	22511	22531	22541	22543
22549	22567	22571	22573	22613	22619	22621	22637	22639	22643
22651	22669	22679	22691	22697	22699	22709	22717	22721	22727
22739	22741	22751	22769	22777	22783	22787	22807	22811	22817

22853	22859	22861	22871	22877	22901	22907	22921	22937	22943
22961	22963	22973	22993	23003	23011	23017	23021	23027	23029
23039	23041	23053	23057	23059	23063	23071	23081	23087	23099
23117	23131	23143	23159	23167	23173	23189	23197	23201	23203
23209	23227	23251	23269	23279	23291	23293	23297	23311	23321
23327	23333	23339	23357	23369	23371	23399	23417	23431	23447
23459	23473	23497	23509	23531	23537	23539	23549	23557	23561
23563	23567	23581	23593	23599	23603	23609	23623	23627	23629
23633	23663	23669	23671	23677	23687	23689	23719	23741	23743
23747	23753	23761	23767	23773	23789	23801	23813	23819	23827
23831	23833	23857	23869	23873	23879	23887	23893	23899	23909
23911	23917	23929	23957	23971	23977	23981	23993	24001	24007
24019	24023	24029	24043	24049	24061	24071	24077	24083	24091
24097	24103	24107	24109	24113	24121	24133	24137	24151	24169
24179	24181	24197	24203	24223	24229	24239	24247	24251	24281
24317	24329	24337	24359	24371	24373	24379	24391	24407	24413
24419	24421	24439	24443	24469	24473	24481	24499	24509	24517
24527	24533	24547	24551	24571	24593	24611	24623	24631	24659
24671	24677	24683	24691	24697	24709	24733	24749	24763	24767
24781	24793	24799	24809	24821	24841	24847	24851	24859	24877
24889	24907	24917	24919	24923	24943	24953	24967	24971	24977
24979	24989	25013	25031	25033	25037	25057	25073	25087	25097
25111	25117	25121	25127	25147	25153	25163	25169	25171	25183
25189	25219	25229	25237	25243	25247	25253	25261	25301	25303
25307	25309	25321	25339	25343	25349	25357	25367	25373	25391
25409	25411	25423	25439	25447	25453	25457	25463	25469	25471
25523	25537	25541	25561	25577	25579	25583	25589	25601	25603
25609	25621	25633	25639	25643	25657	25667	25673	25679	25693
25703	25717	25733	25741	25747	25759	25763	25771	25793	25799
25801	25819	25841	25847	25849	25867	25873	25889	25903	25913
25919	25931	25933	25939	25943	25951	25969	25981	25997	25999
26003	26017	26021	26029	26041	26053	26083	26099	26107	26111
26113	26119	26141	26153	26161	26171	26177	26183	26189	26203
26209	26227	26237	26249	26251	26261	26263	26267	26293	26297
26309	26317	26321	26339	26347	26357	26371	26387	26393	26399
26407	26417	26423	26431	26437	26449	26459	26479	26489	26497
26501	26513	26539	26557	26561	26573	26591	26597	26627	26633
26641	26647	26669	26681	26683	26687	26693	26699	26701	26711
26713	26717	26723	26729	26731	26737	26759	26777	26783	26801
26813	26821	26833	26839	26849	26861	26863	26879	26881	26891
26893	26903	26921	26927	26947	26951	26953	26959	26981	26987
26993	27011	27017	27031	27043	27059	27061	27067	27073	27077
27091	27103	27107	27109	27127	27143	27179	27191	27197	27211
27239	27241	27253	27259	27271	27277	27281	27283	27299	27329

27337 27361 27367 27397 27407 27409 27427 27431 27437 27449
27457 27479 27481 27487 27509 27527 27529 27539 27541 27551
27581 27583 27611 27617 27631 27647 27653 27673 27689 27691
27697 27701 27733 27737 27739 27743 27749 27751 27763 27767
27773 27779 27791 27793 27799 27803 27809 27817 27823 27827
27847 27851 27883 27893 27901 27917 27919 27941 27943 27947
27953 27961 27967 27983 27997 28001 28019 28027 28031 28051
28057 28069 28081 28087 28097 28099 28109 28111 28123 28151
28163 28181 28183 28201 28211 28219 28229 28277 28279 28283
28289 28297 28307 28309 28319 28349 28351 28387 28393 28403
28409 28411 28429 28433 28439 28447 28463 28477 28493 28499
28513 28517 28537 28541 28547 28549 28559 28571 28573 28579
28591 28597 28603 28607 28619 28621 28627 28631 28643 28649
28657 28661 28663 28669 28687 28697 28703 28711 28723 28729
28751 28753 28759 28771 28789 28793 28807 28813 28817 28837
28843 28859 28867 28871 28879 28901 28909 28921 28927 28933
28949 28961 28979 29009 29017 29021 29023 29027 29033 29059
29063 29077 29101 29123 29129 29131 29137 29147 29153 29167
29173 29179 29191 29201 29207 29209 29221 29231 29243 29251
29269 29287 29297 29303 29311 29327 29333 29339 29347 29363
29383 29387 29389 29399 29401 29411 29423 29429 29437 29443
29453 29473 29483 29501 29527 29531 29537 29567 29569 29573
29581 29587 29599 29611 29629 29633 29641 29663 29669 29671
29683 29717 29723 29741 29753 29759 29761 29789 29803 29819
29833 29837 29851 29863 29867 29873 29879 29881 29917 29921
29927 29947 29959 29983 29989 30011 30013 30029 30047 30059
30071 30089 30091 30097 30103 30109 30113 30119 30133 30137
30139 30161 30169 30181 30187 30197 30203 30211 30223 30241
30253 30259 30269 30271 30293 30307 30313 30319 30323 30341
30347 30367 30389 30391 30403 30427 30431 30449 30467 30469
30491 30493 30497 30509 30517 30529 30539 30553 30557 30559
30577 30593 30631 30637 30643 30649 30661 30671 30677 30689
30697 30703 30707 30713 30727 30757 30763 30773 30781 30803
30809 30817 30829 30839 30841 30851 30853 30859 30869 30871
30881 30893 30911 30931 30937 30941 30949 30971 30977 30983
31013 31019 31033 31039 31051 31063 31069 31079 31081 31091
31121 31123 31139 31147 31151 31153 31159 31177 31181 31183
31189 31193 31219 31223 31231 31237 31247 31249 31253 31259
31267 31271 31277 31307 31319 31321 31327 31333 31337 31357
31379 31387 31391 31393 31397 31469 31477 31481 31489 31511
31513 31517 31531 31541 31543 31547 31567 31573 31583 31601
31607

Contributor Index

Below we list those who found (and in some cases just submitted) the *Prime Curios!* in this book. We wish to again thank all of you who contributed to the success of this project.

Subject Index

Many of these terms are defined in the glossary, others are defined in the *Prime Curios!* themselves. The **boldfaced** entries should indicate the key entries.

Slighted Primes Index

When we included a magic square, pyramid, sequence, or any other object made of primes in this book, we always did so listed as an entry under one of the primes in that object. But what of the other primes? Here we list the *slighted primes*, those that occur explicitly in one of the curio entries, but do not have an entry of their own.

Image Credits

Neptune/Watanabe CD p. 22 © John Neptune, used by permission. *Australopithecus afarensis* ("Lucy") 46 by Vincent Mourrem, Creative Commons License 2.5. Arecibo Message p. 55 © Arne Nordmann, used by permission. Photoelectric Sieve p. 67 © The Computer History Museum, Courtesy of The Computer History Museum. Erdös p. 112 © Kilian Heckrodt, used by permission. Edison p. 113, Descartes p. 119, and Franklin p. 140 courtesy of the University of Texas Libraries, The University of Texas at Austin. License Plate p. 159 © Landon Curt Noll, used by permission. USS Prime p. 132 © Paul Wamsley, used by permission. Mersenne Stamp p. 192 and Turing Stamp p. 219 © Jeff Miller used by permission. Lucas p. 232 © Francis Lucas, used by permission.

NGC613 p. 99 Carnegie Institution of Washington, The Carnegie Atlas of Galaxies v. II, Sandage & Bedke, 1994, obtained in digital form from the NASA/IPAC Extragalactic Database, used by permission.

Public domain images. United States coin images pp. 12 and 19 from the United States Mint. Wikipedia: Stegosaurus p. 24, Skewb Diamond p. 174. MacTutor, Univ. of Andrews Scotland: Mersenne p. 23, Eisenstein p. 33, Newton p. 53, Pascal p. 79, Pythagoras p. 89, Euler p. 38, and Gauss p. 120. Parthenon p. 25 Karen J. Hatzigeorgiou. Chemical Structure of Penicillin p. 41 "Cacycle". Chlorophyll p. 70 David Richfield. Unabomber p. 70 Jeanne Boylan, FBI. Nautilus p. 97 Naval Historical Center, U.S. Navy. Fields Medal p. 101 Stefan Zachow. Truck with Trailer p. 49 U.S. Transportion Research Board. Rubik's Cube p. 224 Open Clipart Library.

Sun Card p. 27, Train p. 92, Royal Clock p. 115 and Flush p. 178, © G. L. Honaker, Jr., used by permission. The other images not listed on this page are © C. Caldwell.

Printed in Great Britain
by Amazon